The Environmental Humanities

The Environmental Humanities

A Critical Introduction

Robert S. Emmett and David E. Nye

The MIT Press
Cambridge, Massachusetts
London, England

This book was set in ITC Stone Serif by Jen Jackowitz. Printed and bound in the United States of America.

Library of Congress Cataloging-in-Publication Data

Names: Emmett, Robert S., 1979- author. | Nye, David E., 1946- author.
Title: The environmental humanities : a critical introduction / Robert S. Emmett and David E. Nye.
Description: Cambridge, MA : MIT Press, [2017] | Includes bibliographical references and index.
Identifiers: LCCN 2017003352 | ISBN 9780262036764 (hardcover : alk. paper) | ISBN 9780262534208 (pbk. : alk. paper)
Subjects: LCSH: Environmental sciences. | Science and the humanities. | Education, Humanistic.
Classification: LCC GE40 .E49 2017 | DDC 333.7--dc23 LC record available at https://lccn.loc.gov/2017003352

10 9 8 7 6 5 4 3 2 1

Contents

Acknowledgments vii

1 The Emergence of the Environmental Humanities 1
2 Place, Ecotourism, and the New Wilds 23
3 Energy, Consumption, and Sustainable Cities 47
4 Promises and Dangers of Science 71
5 The Anthropocene's Dark Visions 93
6 Putting the Brakes On: Alternative Practices 117
7 Unsettling the Human 139
8 Conclusions 163

Notes 177
Bibliography 201
Index 217

In memory of Anthony Carrigan (1980–2016)

Acknowledgments

We would like to thank MISTRA (Stiftelsen för miljöstrategisk forskning), a public Swedish foundation that supports environmental research, for the opportunity to explore the emergence and diffusion of the environmental humanities in order to prepare a report (titled The Emergence of the Environmental Humanities) in the spring of 2013. The other consultants on this team were Linda Rugg (University of California, Berkeley) and James Fleming (Colby College, Maine), both of whom contributed many ideas and enriched the result. A reading of that report will quickly reveal, however, that the present volume is a considerably longer and quite different publication. Its form developed through a fruitful dialogue with Beth Clevenger, our editor at MIT Press, whose enthusiasm for the project was infectious and energizing. We are also indebted to five anonymous reviewers, two for the original proposal and three more for the completed manuscript. Our universities provided the libraries and other forms of institutional support we required.

Both authors thank MISTRA's Johan Edman for guidance before this book was thought of. Each of us would like to single out particular individuals for their comments. David E. Nye has learned much from conversations over the years with Martin Melosi, Joel Tarr, Jeff Bolster, Hannes Bergthaller, Eric Sandeen, Klaus Benesch, Torben Larsen, Miles Orvell, Leo Marx, Lawrence Buell, Mark Luccarelli, Steven Hartman, Anna Storm, Sarah Elkind, and members of the Nordic Interdisciplinary Environmental Studies network. None of these good people, of course, should be held accountable for the arguments made here. He also thanks his colleagues at the Center for American Studies for their congeniality and interest in the project. Robert S. Emmett thanks the Rachel Carson Center, its staff and especially Directors Christof Mauch and Helmuth Trischler, who have made the LMU Munich a world center for Environmental Humanities research. Many fellows and visiting speakers provided ideas that found their way into this book. He is particularly grateful for conversations with Marco Armiero, Hal Crimmel, Thomas Lekan, John Meyer, Gregg Mitman, Ursula Muenster, Ruth Oldenziel, Cindy Ott, Ulrike Plath, Jenny Price, Tom Princen and Thom van Dooren. Doctoral students of the Environment and Society program at LMU—especially Nils Hanwahr, Claire Lagier, and Antonia Mehnert—challenged him repeatedly in useful ways. This brief list simply cannot do complete justice to all the good work being undertaken in humanities programs around the world to address environmental problems.

Finally, we thank our families who lived through this project too, at times overhearing our Friday morning discussions over the Internet. They contrived to keep us grounded in everyday life.

Figure 1.1
"The Great Auk." Source: Art and Picture Collection, New York Public Library.

1   The Emergence of the Environmental Humanities

Since the beginning of the twenty-first century a new academic field has emerged: the environmental humanities. This interdisciplinary endeavor developed simultaneously in many parts of the world. It achieved a self-conscious international identity and a name after decades of research by individuals and after the formation of academic associations interested in environmental issues. This book is not an intellectual history of this movement's diverse origins. Rather, it introduces the field to anyone whose interests it intersects. We explain key concepts, central issues, and current research foci. Along the way we also present some exemplary projects. We have tried to avoid partisanship. We want readers to acquire a knowledge of possible approaches and an awareness of debates within the field. We are not seeking converts to a particular method or ideology.

That said, we share some strongly held positions with most people working in the environmental humanities. On the basis of the scientific evidence, we think global warming is taking place, and do not think it is an open question as to whether human beings have contributed to it or not. We believe that species extinction is occurring at an alarming rate, and we reject the notion that humanity has a special place in creation that legitimizes the elimination of other forms of life. We believe that current consumption of the earth's resources is not sustainable. The seas are overfished, the air is increasingly polluted, the oceans contain vast floating islands of plastic, and the amount of garbage produced by human consumption grows year by year. We think that scientists excel at identifying and explaining such problems, but they alone cannot solve them. Solutions will require political and cultural expertise as well. One can build a self-sufficient solar house, for example, but that does not mean the average consumer will buy one. One can design an energy-efficient city, but convincing the public to commit resources to build it and then to live in it is not a scientific problem but an interdisciplinary one.

For example, McKinsey & Company developed a plan for an exemplary ecological community of 50,000 to be built on marshy land near Shanghai. Its first stage was slated for completion in time for the world's exposition held there in 2010. Afterwards, it was expected to grow into a city of a million in three decades. In fact, nothing was built. The local farmers were never consulted, nor were scientists studying rare birds in the area included in the planning process, which was very much top-down. The politicians involved were removed after fraud convictions. As in many such projects, the design imposed had little local input, as was also the case with the well-intended ecovillage of Huangbaiyu in northeastern China. Its failure is a cautionary tale against pursuing technological fixes without adequate insight into the historical and cultural context of a proposed solution. A few brick houses were built in Huangbaiyu, but they did not use the materials specified, and they were poorly designed for local farmers, who said the yards were the wrong shape and size for their needs. Many refused to move into the new houses, whose cost overruns raised prices above what they could afford. A journalist who studied the debacle wrote: "Without extensive consultation with local people, it's a challenge for foreign planners, even with the best of intentions, to understand what is required to transplant a farmer who grew up plowing fields into a city dweller."[1]

Clearly, the humanities have a crucial role to play in understanding and in solving environmental problems such as designing new communities and revitalizing aging cities. However, in this book we do not claim that we already know how to solve the many fundamental challenges to society. We seek, rather, to show how humanists are improving our understanding of the problems and contributing to their solution. That brings us to a final strongly held position across the environmental humanities: that humanists must offer constructive knowledge as well as criticism. Fortunately, this view seems to be widely held by millennials. Surveys show that millennials, far more than members of older generations, give environmental issues high priority and actively seek solutions.

This book surveys key concepts, influential theories, and central debates of a rapidly evolving field. It is an introduction, not a catalog. It moves from earlier theories to more recent ones, and in the process shows that it is impossible to separate environmental analysis from discussions of industrialization and western imperialism, which together accelerated resource extraction, consumption, pollution, population growth, species extinction, and global warming. Even a brief excursion into the literature of the environmental humanities will demonstrate, too, that many striking contributions have come from scholars concerned with gender, race, and queer theory. Our goal is not to adjudicate every debate or champion one theory over another so much as it is to introduce a rich array of ideas in such a way that the reader can gain a preliminary grasp of the environmental humanities and, using the bibliography, decide what to read next. If we think a concept is inherently confusing or a theory flawed, we will say so. However, we have not searched for straw men to demolish; rather, we have focused on what seem the most intriguing and potentially productive approaches.

Emergence of the Environmental Humanities

One can trace the origins of the environmental humanities back more than a century, but the field originated most immediately through the confluence of simultaneous developments during the 1970s and the 1980s in departments of literature, philosophy, history, geography, gender studies, and anthropology. An innovative cluster of Australian researchers adopted the name "ecological humanities" in the late 1990s. They were working at the intersection of history, indigenous studies, anthropology, philosophy, political theory, and nonfiction writing. Simultaneously, at the Massachusetts Institute of Technology, the MacArthur Workshop on Humanistic Studies of the Environment (1991–1995) sought to transcend the common duality of nature vs. culture by locating ecological problems in the behavior of human institutions, beliefs, and practices. Permanent centers that combined natural and social sciences with humanities and encouraged research across disciplines began to emerge during the 1990s.

Such efforts built on a clutch of seminal works. In the United States, Henry David Thoreau's *Walden* (1854) inspired later writers, including John Muir, Jane Addams, Aldo Leopold, Edward Abbey, Kathleen D. Moore, and Terry Tempest Williams, to combine environmental and social criticism.[2] George Perkins Marsh's *Man and Nature* (1864) influenced science writers in the United States from Rachel Carson to Bill McKibben and offered an early model of environmental history. Marsh also represented a European scientific and technical tradition of studying environment and society, often labeled historical geography. It was actively deployed as part of British and French colonialism.[3] A recent anthology titled *The Future of Nature* reflects this aspect of an intellectual genealogy for environmental humanities. To cite only a sample of three potent ideas from its collection: Alexander von Humboldt and Aimé Bonpland's advocacy in *Essay on the Geography of Plants* (1807) for an integrated, planetary study of nature; William Stanley Jevons's writing on the question of coal supply, consumption, and the British Empire (1865); and Vladimir Vernadsky's concept of the biosphere (1926).[4] There is a direct line from such thinkers to the development of the environmental humanities, which consider the environment and humanity to be inextricably connected and refuse to preserve an unproductive hierarchy among forms of knowledge. It is imperative to abandon narrow disciplinary traditions in order to grasp these interconnections.

The environmental humanities did not simply evolve from earlier Western thinkers, however. In fundamental ways, the field has been shaped by postcolonial and feminist studies and by scholars working outside of Europe and the United States. These scholars critiqued the limited focus of earlier environmental histories and the literary canon of "nature writing." They challenged the persistent anthropocentric (and at times imperial) bias of dominant ethical and political theories. They pointed out that the very idea of humans standing outside of and controlling nature was based on a conception of the natural world that was passive and feminized. There was often an implicitly racialized conception of the natural world that depended on a restricted, historical *Anthropos* that was conceived as white, male, and European. Apparently neutral terms such as "man and nature," though they still appear in policy discussions of climate change, often have masked unequal social relations and exploitation of resources. In a major shift of perspective, a new range of concepts emerged that provide a framework for environmental humanities, such as ecoracism, environmental justice, "naturecultures," the environmentalism of the poor, and the posthuman.

The environmental humanities has become a global intellectual movement that reconceives the relationship between scientific and technical disciplines and the humanities, which are essential to understanding and resolving dilemmas that have been created by industrial society. By 2010, scholars in Australia, North America, and Western Europe had begun to embrace the term "environmental humanities," and today scholars with similar interests can be found on every continent. Yet half a century ago several fields that have contributed to the environmental humanities— notably literature, cultural geography, anthropology, and history—had already begun to bridge the gaps between themselves and science.

The environmental humanities provide historical perspectives on the natural and social sciences, pointing to how their agendas, initial goals, and occasional failures have been affected by political ideologies and economic interests. The environmental humanities also assist in the interpretation of scientific results and technical innovations. Even multidisciplinary scientific commissions sometimes produce knowledge that is overlooked or underused. For example, the historian Tom Griffiths has described how a Royal Commission in Australia compiled thousands of pages on the damaging effects of pastoralism and overstocking on the continent more than a hundred years ago. Australians in power have long had the knowledge of ecological limits. Griffiths cites this as a prime reason we need ecological humanities, concluding: "Scientists often argue for the need to overcome deficits of knowledge, but rarely ask why we do not act upon what we already know. Most of the constraints working against environmental change are cultural: we have to know ourselves as well as the country."[5] Moreover, those who discover or invent often prove poor prophets when seeking to explain how their discoveries will be used or abused. The researchers and managers who invented the Internet, which now is widely acknowledged as a planetary cultural force, hardly grasped its far-reaching potential. The American Telegraph and Telephone Company turned down a chance to purchase the nascent technology.[6] Nor is this an isolated example of the inability to forecast the future uses of an invention.[7]

There is little reason to believe that scientists who improve cloning techniques, design solar radiation management systems, or manipulate DNA to create new forms of life are able to see all the possible results of their work, either. The environmental humanities can address ethical problems that will arise as cloning becomes more widespread, or as governments and private individuals decide whether to spray sulfur into the upper atmosphere to combat a "climate emergency," or as the courts litigate questions of the ownership of DNA, or as international agencies wrestle with problems of digital property rights. The environmental humanities are clarifying the origins and consequences of social practices that are involved when a government or community wants to change energy consumption or resource use. They are identifying the implicit narratives about energy and natural resources that shape public opinion and suggesting new narratives that can help people to understand and embrace new practices.[8] They have begun to foster the values needed to build an environmentally responsible society with habits of ecological citizenship. Carolyn Merchant has argued that "a partnership ethic would bring humans and non-human nature into a dynamically balanced, more nearly equal relationship." Such a new ethic requires a new narrative. This new story "would not accept the idea of subduing the earth, or even dressing and keeping the garden, since both entail total domestication and control by human beings. Instead, each earthly place would be a home, or community, to be shared with other living and non-living things."[9] In such efforts, scholars and writers are breaking down academic barriers between the humanities and the sciences, even as these separations are being breached in the larger society.

Because of its wide range, the field of environmental humanities is difficult to pin down, and it has different profiles depending on the scholarly strengths at the institutions where it has emerged. One university may have a strong environmental history group in dialogue with groups specializing in postcolonial studies and anthropology. Its faculty might edit a journal focused on ecological imperialism. Another university may be weak in these areas but foster a productive dialogue between scholars in ecocriticism, environmental ethics, and gender studies. Their publications might refer only occasionally to ecological imperialism but build a robust dialogue on the dimensions of the human, focusing on transcorporeality, vital matter, and transgender studies. With such differences between research groups, the field might appear incoherent, but it is more accurate to say that it evolving rapidly.

Ten years from now, the environmental humanities will likely be present in most universities. This prediction is based on the fact that the institutions with such programs are among the world's elite. In the United States the list includes Stanford University, the University of California at Los Angeles, Princeton University, the University of Pennsylvania, and the University of Wisconsin. But the United States is not necessarily in the lead. Important centers also have emerged at leading universities in Britain, Sweden, Germany, Australia, and China. Financial support that once was episodic and haphazard has become more systematic. The environmental humanities are now articulating concerns relevant to medicine, animal rights, neurobiology, race and gender studies, urban planning, climate science, and digital technology. More generally, humanists with environmental concerns are proving adept at communicating with a broad public, evoking a sense of wonder at the resilience and complexity of ecological systems, empathy for endangered species or habitats, and understanding of the urgent need to take knowledge-driven action on environmental issues such as global warming, habitat loss, water pollution, and food insecurity.

The open-minded, constructive approach of the environmental humanities can motivate creative cooperation between the humanities and the sciences and can assist in the interpretation of scientific results. They share a belief in the power of arts and humanities to spark innovations relevant to other disciplines and practical fields. Arts and design advocates have proposed that the acronym STEM be expanded to STEAM to recognize how the arts contribute to inventiveness in science, technology, engineering, and mathematics.[10] Transdisciplinary projects in the environmental humanities have involved storytelling, semi-structured interviews, and visual ethnography to develop usable models for directing energy development, agricultural practices, land use, and water management. For example, state agencies and public museums have employed historians, writers, photographers, and artists to communicate the complex relationships of societies and their environment. These efforts call for craft and training. Initiatives in the environmental humanities are often inspired by artists, filmmakers, playwrights, and specialists in digital media.

The global environmental crisis demands new ways of thinking and new communities that produce environmental solutions as a form of civic knowledge. The crisis cannot be addressed solely by finding technological solutions to particular problems that are delivered "downstream" to a population of passive consumers. The crisis has been caused by human behavior and by institutions that externalize environmental costs and cordon off regimes of violent extraction and waste from view. Relying for solutions on the scientific and engineering specialists who served these very institutions and vested interests seems a maladaptive response. For cultural and political reasons, even the best science and the best technologies—for example, those used to mitigate climate change—are often not adopted. The planetary crisis can best be addressed through an interdisciplinary approach to environmental change that includes the humanities, the arts, and the sciences.[11] Libby Robin recorded a striking remark by an Australian scientific manager for a coastal zone: "'We do not manage the environment,' only the human behaviors that affect its structure and processes."[12] Human beings are not merely observers, they are an active part of nature. Yet the public does not necessarily understand or believe a scientific finding, such as global warming. The humanistic disciplines can help to explain such findings and discover ways to address the public more effectively. Major global financial and development agencies now recognize that addressing the public requires value systems and registries of information that are more nuanced than conventional data such as GDP or CO_2 emissions. They demand broad thinking, teamwork across the disciplines, and knowledge that is *affective*, or emotionally potent, in order to be *effective*, or capable of mobilizing social adaptation.

Some Central Concepts

The critical agenda of the environmental humanities emerged in response to a multi-pronged crisis of ecology, economy, politics, and epistemology. The rumblings of a global ecological crisis, first widely heard in the 1970s, inspired a raft of monographs, "green" special issues of journals, and conferences. At that time there was virtually no awareness of global warming, but a good deal of discussion of acid rain from pollution, the population explosion, the rapid consumption of the earth's resources, and the possibility that even a "small" nuclear war could stir up so much dust in the atmosphere that the earth would endure a "nuclear winter" that could wipe out most of humanity. One of the most widely discussed topics was that of "the limits to growth." This initial surge of interest receded somewhat in the 1980s, but history and literature had created sub-fields in environmental history and ecocriticism, with similar developments in other departments. These separate initiatives began to coalesce by the 1990s. The growing list of environmental challenges—acid rain, species extinction, genetic engineering, global warming—further stimulated a new generation of scholars.

As institutions grew, methodological sophistication and professional credibility strengthened and scholars began to ask penetrating qualitative questions about the human-environment nexus. Membership (and cross-membership) in key organizations for environmental history, ecocriticism, philosophy, anthropology, geography, feminist theory, and postcolonial studies grew dramatically after 2000. These changes came on the heels of the political and epistemological struggles of the 1990s, when many wrote of "the humanities in crisis."[13] Higher education systems faced financial restructuring, which spurred public debate about the future of research in the humanities. A critical mass of researchers and teachers began to propose the environmental humanities as a positive response to complex social-environmental problems. Many of them had strong bases of experience and support in public service groups, museums, and nonprofit organizations. The practical orientation of their work reflected a new confidence in the value and organization of humanistic research. Leading voices called for a robust role for the humanities alongside the social and natural sciences to address longer-term challenges that will not yield to quick technical fixes.[14]

A few central ideas anticipated the environmental humanities from within the disciplines that contributed to their emergence. For example, some environmentally minded philosophers rejected the idea of "the thing in itself" (Immanuel Kant's *das Ding an sich*) in a new way.[15] Kant wrote of the difficulty of apprehending any object through the senses. But environmental philosophy asserted that a disconnected and isolated "thing" or object does not and cannot exist. Rather, every object and being is defined by its relationships. It is part of networks and only has meaning in relation to its surroundings. Scientists studying nature had sought to know the "thing in itself" and to isolate an organism in a cage or a glass container. Yet to grasp fully any form of life requires studying it in its habitat, where its existence is defined by relations with others of the same species, and by the plants, animals, insects, and microscopic organisms that share its environment. As the analysis becomes more detailed, it must include more and more about the environment, including the climate, food sources, predators, competitors, procreation, and so on. Less knowledge can be attained by studying "the ant itself" (which will soon die if kept isolated) than by studying an anthill in its normal environment. Studying an ant colony traces larger patterns in space, following ants as they inhabit their surroundings and respond to the round of the seasons. A few critics of Kant went further and agreed with Alan Watts, an interpreter of Eastern religions, who declared that "what we call things are no more than glimpses of a unified process."[16]

Human beings are not independent of the natural world, but are part of it. From an ecological perspective, the nature/culture dichotomy that was common during much of the nineteenth and twentieth centuries makes

no sense, though this dualism remains one of the most powerful conceptions of nature.[17] Human beings are animals and are mortal. Two quite
different groups do not agree with this proposition, however. On the one
hand, some conservative Christians reject the theory of evolution, and see
human beings based on Biblical authority not as animals but as beings
made in the image of God. Indeed, there is even a creationist museum
that purports to demonstrate the historical accuracy of the Biblical creation
story.[18] From this perspective, human mortality is the prelude to an immor-
tal afterlife. Scholars steeped in Enlightenment rationality reject this view.
Instead, they see human beings as being by their very nature embedded in
the natural world. From this perspective, religion provides no escape route
from mortality.

Yet Christianity and other religions are not necessarily at odds with the
environmental humanities. Indeed, religious leaders and theologians are
well represented in recent anthologies on environmental ethics and climate justice.[19] Pope Francis echoed radical social ecologists with his call
in the 2015 encyclical *Laudato Si'* for humans to respond to the planetary
crises of social inequality and ecological degradation.[20] Pope Francis also
proclaimed St. Francis of Assisi "the patron saint of all who study and work
in the area of ecology." There have been both Catholic and Protestant theologians with a decidedly environmentalist view of the world. To develop
such a perspective requires an acceptance of our biological mortality and
our dependence on other species.

From an entirely different perspective, a smaller group believes that
human beings are about to evolve beyond mortality. They think that
human beings and machines will merge by combining advances in biology
and computing, and they argue that perpetual life is possible for mankind.[21]
Most scholars in the environmental humanities are not comfortable with
such views, which are advanced by some scientists and futurologists. One
of the editors of the magazine *Wired*, Kevin Kelly, argues that technology is
not cultural but natural, and that it is a part of an evolutionary process that
accelerated 50,000 years ago when human beings invented language. Kelly
is convinced that technology is a manifestation of natural selection, that it
is directional, and that it "wants" to achieve more efficiency, more opportunity, and more complexity, diversity, specialization, freedom, beauty, sentience, mutualism, ubiquity, structure, and "evolvability." Human beings,
in this perspective, are unfinished, and their development is accelerating.

Some futurologists predict the emergence of superintelligent machines within a generation. Such views perpetuate the division between human beings and nature, the latter understood as raw material to be mined and manipulated so that human beings can evolve into genetically enhanced cyborgs. This perspective puts human beings firmly at the center and sees little inherent value in other species. Kelly declares, for example, that "we can see more of God in a cell phone than a tree frog."[22] But most cell phones are on the scrap heap within two years of their manufacture, whereas tree frogs, which are disappearing at an alarming rate, survived millions of years before human beings began to undermine their ecologies.

Human beings do not have special rights relative to other species. Rather than view animals and plants in terms of their usefulness to humanity, we can see them as having an intrinsic right to exist. And when an entire form of life disappears, its loss diminishes human culture too. As Thom van Dooren writes, we must move beyond a "core of a human exceptionalism that holds us apart from the rest of the world and, as such, contributes to our inability to be affected by the incredible loss of this period of extinctions, and so to mourn the ongoing deaths of species."[23] These fundamental ideas of intrinsic value, reciprocity, and a right to exist are salient in studies of interspecies relationships, but they are also foundational for other scholars. In Judeo-Christian religious terms, this view is at odds with a literal interpretation of the Old Testament. Yet it is also possible to understand the Garden of Eden as an ecological system that Adam and Eve were responsible for, rather than to view them as landowners with an implicit right to kill what they pleased. The Sioux espoused ideas compatible with the view that all species have rights, as many Buddhists believe. The Sioux leader Black Elk understood the interdependence of his people and the ecology of the Great Plains, and could not understand why white society was determined to slaughter enormous herds of buffalo and take only their hides.[24] At a minimum, bison had a right to be used properly and to have their spirit acknowledged when they were taken by hunters.

The environmental humanities reject the notion that Western cultures are superior to other cultures, and recognize that much knowledge is place-specific. During the age of European imperial expansion (roughly from the time of Columbus until the middle of the twentieth century), technological superiority was often taken to be proof of cultural superiority. Agricultural "improvement" of land and then control of advanced machines became the measures of cultural development.[25] Reprehensible as this view may seem today, it at least assumed that no people were inherently inferior, and that a people might become "advanced" by adopting Western technologies. However, the idea that more complex and powerful technologies are inherently better is difficult to maintain. A high level of science and engineering might seem a proof of superiority, yet no country had a more advanced scientific community than Germany when the Nazi Party took control. Is North Korea superior to South Korea because it has atomic weapons? Were the Spanish culturally superior to the Aztecs because they had steel swords and armor and used the wheel in transportation?

The environmental humanities do not seek to establish a hierarchy of cultures, as was explicitly the case in many disciplines as they defined themselves from the middle of the nineteenth century until after World War II. A cultural hierarchy was quite explicit in early anthropology, with its evolutionary typology from hunter-gatherers to "advanced" industrial peoples, and was foundational in a wide range of other disciplines, including political science, the history of art, and the natural sciences. Whereas a century ago the dominant assumption was that all cultures ought to become "developed" on a Western model, the environmental humanities recognize that not all cultures are following the same historical trajectory. World cultures are not, nor should one try to make them, homogeneous. Rather, cultures often focus on local ecological knowledge and express how particular groups live within each place. Every culture is distinctive in part because it evolves within specific locations. The patterns of life appropriate on the coasts of Greenland or on the plains of Tanzania are not transferable to Berlin or Los Angeles, or vice versa. To put this another way, every culture is constantly evolving and adapting, and each finds its own path.

Nevertheless, every society must deal with the global problem of accelerating growth. What are our different freedoms, rights, and responsibilities in an age of extinctions, resource limits, and climate instability? A new terminology has emerged that frames ecological issues in new ways. Brief definitions of some of these interrelated ideas follow.

We begin with the *crisis of the commons*, sometimes also called the *tragedy of the commons*. In many societies there are some lands that are owned by a state or a community rather than by individuals, and some shared activities take place there—for example, hunting or grazing domestic animals. In earlier times many societies held most or even all of their land in common, and this is still the case for some aboriginal peoples. Collective land owner-ship was common in Britain in pre-Roman times, and it remained impor-tant until the feudal system emerged, with written documents that both registered property ownership and also protected local rights to use com-mon lands. Britain still has such lands, which despite what the name seems to imply are not available to everyone. Rather, only local residents have the right to use common lands, for example, to graze a horse or cow.[26] Fishing grounds provide another example of an environment held in common but which may be overexploited if there are inadequate safeguards to ensure it remains viable. When such sites were shared locally, the community under-stood the need to conserve resources for the long term. In recent times, for a variety of reasons, fewer people have taken responsibility for such spaces. This may be because outsiders begin to use the commons, as has happened with many fishing grounds. But the failure to take responsibility is often an internal problem. Whether one blames this failing on increased personal mobility that weakens the sense of being part of a local commu-nity, on the overuse of shared scarce resources by greedy individuals, on the lack of environmental controls to prevent pollution, on the carelessness of industries discarding wastes, on legal enclosures of commons to privatize economic benefits, or on other factors, many sites held in common have become severely compromised. They no longer are a shared resource that brings a community together. Debate over the causes and significance of what came to be called the "tragedy of the commons," after the title of an article by the ecologist Garret Hardin published in 1968—a tendency of common pool resources to be overused—rages on.[27] This "crisis of the com-mons" had clearly emerged as an issue by the end of the 1970s, when the public, journalists, and scholars agreed on that name for it.

The problem was not new, but historically it could be seen in several ways. In Britain, for example, common lands had been subjected to enclo-sure acts that allowed landowners to acquire them. Similar acts were part of the encounter between First Peoples and imperial expansion. The term "commons" itself presaged the development of the environmental humani-ties, and several additional terms were developed from this one, including "commoning," "transition," and "re-skilling." One lesson of the commons is that in "conservation, human management is vital." Rather than expel indigenous people to create national parks and nature preserves, as has been done in Africa, the idea of the commons "includes both people and other species" in a sustainable relationship.[28] The commons can be digital as well as physical space, with such collaborative projects as Wikipedia exemplifying the possibility that sharing resources can include knowledge as well as pasture land or forests. One popular conception of the Internet—an idea championed by the Electronic Frontier Foundation—was as a cyberspace that was outside markets and national boundaries.[29]

Thinking about the crisis of the commons became intertwined with the realization that people, particularly those in advanced industrial societies, had too little respect for other life forms. Rather than see history solely in human terms, we should include other life forms, seeing them as entangled with humanity. In trajectories of co-evolution, human beings have emerged alongside *companion species*. In the fields, the shepherd, his dogs, and the sheep mutually define one another. The rhythms of a dairy farmer's day express his relationship to the herd. Falconers learn to understand the way different birds of prey behave, and know, for example, how falcons differ from hawks. As one experienced woman put it, "Each one of the raptors I have handled has proven to be a unique individual."[30] Philosophers and anthropologists write of human and animal *co-becoming*, whether the animals in question are wild species we have struggled to preserve or domesticated cats, dogs, horses, cows, and birds.[31] Good farmers can also take care of wild species. In Tanzania one initiative of the Jane Goodall Institute is to plant shade grown coffee in areas near chimpanzee reserves. The chimps show no interest in the beans, and their tree cover is preserved.[32] The human-animal relationship is understandably at the heart of much story telling. As ecocritics, semioticians, and scholars of religion have shown, myths, symbols, and stories of potent animal others—a coyote trickster, an ancestral Kangaroo, or Odin's raven—guide human choices and flesh out life's meaning.

But the practices of co-becoming are fundamentally threatened. One of the most widely read books of 2014 was Elizabeth Kolbert's *The Sixth Extinction*.[33] Although the idea of the sixth extinction was not original with her, Kolbert—a skilled writer for *The New Yorker*—spread it to a much larger audience. She began by describing five extinctions that occurred before the emergence of mankind, notably the one that killed off the dinosaurs. These five extinctions were due to catastrophic natural events, such as a meteor's striking the earth and throwing millions of tons of dust into the atmosphere and resulting in drastic cooling of the planet. The sixth extinction, which has continued with little interruption since human beings emerged in Africa, has not been caused by a natural disaster, but rather by us.

In most cases, extinction of any one animal or organism was not due to an intentional campaign of eradication but rather was an outcome of uncoordinated, individual actions that changed habitats, climates, and ecological systems. Hunters did not self-consciously set out to wipe out the Great Auk, a large flightless bird once found by the millions from Italy to Newfoundland. Great Auks were once common in Iceland, where a hunter (apparently not aware that he was doing so) killed the last one in the world in 1821.[34] The extinction of the Great Auk required several thousand years. The sixth extinction as a whole has not occurred at a slow and steady rate, but has accelerated. For example, there are extensive coral reefs along the world's coasts, and a century ago it would have seemed unlikely that coral could become an endangered species. But owing to a variety of human interventions, 80 percent of the coral reef cover in the Caribbean has disappeared, and coral could become extinct there.[35] As we prepared the manuscript of this book, many parts of the Great Barrier Reef experienced catastrophic bleaching as a result of sustained high temperatures. In such cases, there are multiple causes that are complexly interrelated.

One reason some species have disappeared is that human beings have assisted invasive species to spread in a process of *ecological imperialism.* Microbes, animals, and plants function as co-invaders that may impose change even more irreversible than that imposed by militarized human colonization. Alfred W. Crosby traced what he called "the biological expansion of Europe" from the Vikings to 1900[36] and documented how European crops replaced native grasses and European domestic animals displaced native species. Weeds such as stinging nettles and dandelions inadvertently landed in European colonies too, as did rats. Other creatures, brought intentionally, escaped and flourished in the wild; among them were honeybees, feral horses in the American West, wild camels in the Australian outback, and millions of cattle on the plains of Argentina. These European species marginalized or eliminated native plants and animals.

On the reverse journey, biota from the Americas, Caribbean, South Asia, and Africa spread through Europe as novel commodities. Ecological imperialism included the search for new commodity crops, some of which supported the spread of plantation agriculture by feeding enslaved laborers (maize, breadfruit, sugar cane), some of which fed industrial workers in

Europe (potatoes, corn, refined sugar), and some of which resulted in the development of new markets and social practices (for example, around tea, coffee, and chocolate). France, Great Britain, and Holland exported plantation crops, cultivation techniques, irrigation technology, and models of management across their colonies. Local resistance to ecological imperialism took many forms—for example, farmers in colonial Egypt intentionally neglected drainage systems, and enslaved Africans smuggled rice and other plants and ran away to found villages in Amazonia, feeding themselves with a hybrid diet of Amerindian and African crops. And later in the twentieth century, many native peoples resisted expropriation of their forests for timber or enclosure for national parks.

Human actions have changed the air, the soil, the seas, and the weather to such an extent that it has been common to speak of the present age as the *Anthropocene*. The Anthropocene began at the moment when human effects on the environment became so great that they registered in the geological record, whether as pesticides in sedimentary layers, as the disappearance of species from the fossil record, as rising sea levels, or as debris orbiting the earth. The effects of humanity are measurable scientifically and have enormous long-term implications. The concept of the Anthropocene places the environmental humanities in a framework of time and space that begins before human beings existed and asks how long they will survive. As interest in the Anthropocene increased, so did research on the history of climate change and on varying understandings of the relationship between climate and local cultures. A six-volume collection of papers edited by Mike Hulme pulled much of this work together, showing that climates are not only physical but also ideological. The concept of climate is also unstable, and it is being constantly redefined.[37]

The term "Anthropocene" expresses a new form of that relation. Before astronomers discovered that the sun is not the center of the universe and geologists discovered that human beings had emerged quite recently in the earth's history, mankind seemed to be at the center of existence. The concept of the Anthropocene curiously restores the importance of human beings, but only as a disruptive force on one small planet in a vast universe.

The Anthropocene raises issues that are beyond the control of individuals or communities. In contrast, the concept of *climate justice* focuses on the local refraction of global forces and their intergenerational effects. It emerged in parallel with the rise of environmentalisms of the poor and the struggles of native peoples' worldwide and with the rise of the *environmental justice* movement in the United States. These largely grassroots movements initially focused on struggles of groups dependent on ecosystems for subsistence and subsequently branched out to include resistance against pollution and hazardous wastes. The latter disproportionately affect the poor and ethnic minorities, while expropriation of common resources, such as forests, water, and game, has been a universal feature of economic neo-imperialism from the Himalayan region of India to Kenya and Tanzania to Brazil and Bolivia. In the United States, the environmental justice movement long regarded mainstream environmentalists as "white, often male, middle- and upper-class, primarily concerned with wilderness preservation and conservation, and insensitive to-or at least ill-equipped" to deal with their problems.[38] All too often, toxic waste sites were located near communities of African Americans, the poor, or other disadvantaged groups, notably along the Mississippi River in Louisiana. There, an area with many petrochemical plants became known as "cancer alley." Houston had similar problems, because its oil industry long faced ineffective environmental regulation.[39] In a 1995 address to the American Society for Environmental History, Martin Melosi recognized that the environmental justice movement had its roots in the civil rights struggle, and that "the Environmental Justice Movement found its strength at the grassroots, especially among low-income people of color who faced serious environmental threats from hazardous wastes and other toxic material." "Women," he continued, "have been key leaders in the anti-toxics effort, including Virginia civil-rights activist Cora Tucker, Lois Marie Gibbs of the Citizen's Clearinghouse for Hazardous Wastes, and Sue Greer, organizer of People Against Hazardous Waste Landfill Sites."[40] What Joni Adamson and Slovic Scott wrote in 2009 holds true for the environmental humanities as a whole: "Environmental justice has provided an effective rubric for the explicit engagement with environmental issues in relation to ethnicity, social class, and gender, but this work is hardly finished."[41] Environmental justice has become a central concern in transnational networks of scholars and activists. For example, in the Philippines a grassroots movement against copper mining on Mindanao thwarted the plans of the Australian Western Mining Corporation and forced it to withdraw from a major project. Since the national government sided with the mining company, it was crucial that the Catholic Church sided with the activists.[42]

Yet extraction of raw materials is by no means the only problem. As pressure to protect the environment intensified in the United States and in the European Union, banned chemicals and wastes began to be outsourced to poor countries, but sometimes returned to the dinner table through imported produce. By 2006 the US was generating 2.6 million tons of "e waste" from discarded computer screens, mobile phones, and the like per year. These contain toxic substances, including lead, and much of the waste was exported to China, India, and Africa.[43] The US General Accounting Office found that the waste was often exported illegally and that disassembly methods abroad were often unsafe, including open fires and acid baths.[44] In short, wealthy countries outsourced their pollution. As this example suggests, some societies have more historical responsibility for scale of ecological damage than others. Likewise, certain groups are more severely affected—for example, because they live on islands that will soon sink beneath rising tides or because their farmlands are drying up. The impulse to examine environmental justice has often come from the so-called Third World, from sacrifice zones within "developed" countries, and from the ragged edge of resource frontiers. Many writers on this subject argue that any sense of ecological justice has to include not only the species alive today (including *Homo sapiens*) but also future generations.

Ecoracism is closely related to the idea of environmental injustice. As late as the 1990s race was not widely discussed as a part of environmental history, but this has changed through dialogue with postcolonial criticism, and engagement with indigenous, anti-globalization and environmental justice movements. For example, it quickly became evident to people studying the location of heavily polluted sites that poor people and people of color disproportionately lived near many of the most severely affected areas. At the other extreme, the notion of a pure, untouched wilderness implied that no human beings had ever lived there. Racist ideologies of white Europeans' supremacy abetted theft of lands, murder of indigenous peoples, and forced relocation to concentrated settlements, often on the most marginal lands, from Australia to South Africa to the Great Plains. In turn, the consequence of economic and cultural damage to indigenous communities reinforced settlers' false sense of racial superiority. Thus, it seemed logically necessary, when creating American national parks in the nineteenth century, to remove Native Americans from sites such as Yosemite and Yellowstone. In the late twentieth century, European conservationists in Africa often pursued a similar logic of exclusion on behalf of wildlife preservation; native African subsistence hunting was blamed for dwindling numbers of megafauna, even as white European hunters killed lions and rhinos for sport. Nor is ecoracism merely "bad history." In 2016, more than 300 indigenous tribes joined the Standing Rock Sioux to protest an oil pipeline that they do not want to pass their reservation and that would pollute their drinking water if it ruptured. The Standing Rock Sioux have been attacked with dogs, pepper spray, concussion grenades and tear gas while protesting as bulldozers cleared a pipeline route adjacent to their sacred burial sites.

Sudden violence, such as an erupting volcano, a burning oil spill, or a clash between protesters and police in riot gear, makes headlines. Other forms of violence occur slowly over years or even decades—the resettlement of ethnic minorities, the poisoning of groundwater—and these often are ignored not only by the media but even by nearby communities. Rob Nixon uses the term "slow violence" to describe how a process such as deforestation or the exposure of villages to industrial pollutants gradually robs a community of sustenance.[45] Repeated oil spills have slowly destroyed the habitat of the Niger River Delta. China's development policies are transforming Tibet and driving out its nomadic peoples.[46] Pesticides are killing millions of bees, which are crucial to plant pollination. Fracking for oil and natural gas has destroyed local aquifers, leaving water supplies so polluted that they are dangerous to health. Irrigation projects in the American West covered millions of acres with water, and gradually polluted them because the water contained selenium, boron, and mercury which concentrated in the fields during decades of evaporation. Nixon's work is part of a general shift away from a concern for nature in the abstract to a focus on endangered human communities in particular settings. It exemplifies the benefits of dialogue between environmental and postcolonial literary studies. This development has also registered in literature—notably in Amitav Ghosh's novel, *The Hungry Tide*, which takes place in an archipelago off the coast of Bengal. A Western scientist cannot complete her field research without a fisherman's detailed local knowledge, and the environmentalism of urbanites intent on saving the tiger from extinction is contrasted with the complex place of the tiger in local culture.[47]

How many people can the earth support? This is a complex question, depending on the average level of energy use, the amount of meat consumed per person, and much else. If everyone is to consume at the German level, much less the American level, the current population is already too large. In the 1970s such thinking was widespread, but in recent decades it has become less visible. Instead, the hope has become that human beings can reduce their "carbon footprint" by recycling, eating less meat, adopting more energy-efficient appliances, and so on. Yet if the world's population keeps increasing, these practices will no longer be sustainable, and an active program of population restriction or reduction may become necessary. *Degrowth* (in French *décroissance*, in Spanish *decrecimiento*) is a concept developed by anti-consumerist thinkers who argue for downscaling production and stopping the (over)use of resources. This movement (discussed further in chapter 6) challenges the common perception that growth is the measure of economic well-being, and argues that human beings are happier in a less competitive society.

Most species in a stable ecological system bounce back from diseases, droughts, or attacks. They are resilient, recovering their numbers and retaining a role in the system as a whole. As a general idea, this is not hard to grasp. But how long does it take a plant population to recover if sprayed with an herbicide? How long (if ever) before a population of cod can recover after being fished to near extinction? And should one idealize their earlier condition and assume that the goal must be to restore an ecological status quo? In a classic article published in 1973, C. S. Holling defined resilience as a "measure of the persistence of systems and of their ability to absorb change and disturbance and still maintain the same relationships between populations or state variables."[48] Holling was seeking an alternative to the concept of sustainability. His idea of resilience built on the assumption that systems constantly change, and that a system that has been disrupted may not return to a previous status quo but instead may attain a different configuration of stability. For example, salmon once threatened with extinction show a capacity for adaptation and manage to survive, though not necessarily in all the habitats they once occupied.

Resilience is a bit slippery as a concept, but it suggests that a healthy ecological system is self-regulating and is able to adapt to external challenges within certain limits. A system, rather than being a balanced arrangement that has persisted for millennia, is always in a process of change. When this idea is applied to a single species, research often concerns its resilience. Is it improving, declining, or threatened? The concept of resilience has been taken up in many fields, including family sociology, archeology, and geography, and a journal explicitly devoted to the environmental humanities is titled *Resilience*.[49] The editors of that journal note that Hollings rejected the idea of "equilibrium as the core of the ecosystem concept in favor of destabilization, in the process rejecting the Club of Rome's *Limits to Growth* and other arguments for sustainable development." They conceive the journal as "a commons in which anyone interested in the humanities and the environment can participate in an evolving conversation."[50]

The terms that have been developed in the environmental humanities are interconnected. Together they suggest a perspective on the world that is fundamentally different than that cultivated by any one discipline. Anyone studying animals as companion species, for example, can see connections between Indian elephants, black rhinos, or wolves and research on de-growth or the crisis of the commons. Likewise, slow violence is often present in the same situations in which one encounters dehumanization and ecoracism. The crisis of the commons exemplifies on a local scale the environmental crisis as a whole. If one takes seriously this range of concepts and the urgent problems that they address, it seems irresponsible to adopt the old-style humanities, working within a single discipline, content to focus on narrowly defined concerns. Instead, the philosopher finds it necessary to think about history, the historian must engage with anthropology, the anthropologist (already in an interdisciplinary field) must engage with technological history, and so on, across the humanities. The walls between departments need to be torn down in order to confront environmental crises. New kinds of focal points have emerged. Building on the interconnections between these concepts, the following chapter will examine place, ecotourism, and wilderness.

Figure 2.1
Ansel Adams, "The Grand Canyon," 1941–1942. Photographs of National Parks and Monuments, 519894, National Archives.

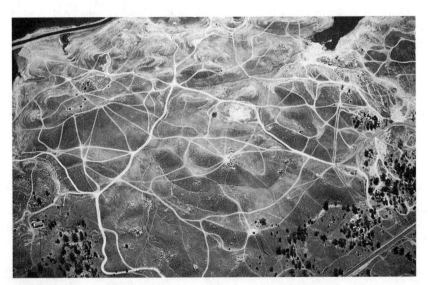

Figure 2.2
Gene Daniels, "Motorcycle tracks along Isabella lakeshore, Kern River area," 1972. Documerica Series, Record Group 412, 512514 National Archives.

2 Place, Ecotourism, and the New Wilds

Shared Sense of Place

Place is one of the most complex yet seemingly simple ideas in the humanities. A closely related idea is that of "place-making." A critical understanding of "place," one not synonymous with an ideological commitment to particular places or rigid localism, is indispensable to rethinking the human relationship to the world. For centuries, Western surveying practices and property markets have encouraged people to think of land as a commodity—as an investment to be manipulated for profit. Land, in this view, is raw material, to be mined or farmed or built on to maximize profits. The time horizon in Western land use is rarely more than the length of a mortgage. And when land is sold, a new owner may bulldoze the site, erasing traces of previous use. Treating land as a commodity implicitly says that human beings stand outside nature.

This attitude toward land has hardly been constant in human history or across cultures. In fact, it is rather recent. To survive as a species, human beings needed an intimate understanding of each location in which they had lived, and for millennia this knowledge was an important part of the heritage passed down from one generation to the next. In one of the cliff dwellings at Mesa Verde in Colorado there is a mark on the rock that the rays of the sun had to reach each spring before planting would commence. A great many such pieces of information, woven together, enabled a community to survive. And note that it was a matter of community. Groups of people inhabited land together, and in most early societies individuals did not own land; ownership was a collective responsibility. There was, in effect, only the commons. For Native Americans on the Great Plains, this meant that hunting was not individualistic but collective. A single hunter did not disturb a herd of buffalo, for example. Rather, the tribe made a coordinated plan, and afterwards the meat was not the property of the hunters alone but was distributed to everyone in the tribe. One must not romanticize and assume that aboriginal peoples never allowed their live-stock to overgraze, never deforested, and never hunted a species to extinc-tion. But on the whole they trod more lightly on the land than modern industrial societies, and most so-called primitive (that is, non-Western or non-capitalist) societies imagined humanity as part of nature rather than as standing outside it.

A psychological identification between self and site is part of place-mak-ing, in an ongoing social process. It demands a personal investment in a particular location whose appearance, sounds, and smells become part of a daily round. Creating a sense of place in an unfamiliar space establishes somewhere to belong. Human geographers and other researchers consider this process of identification to be fundamental to place. Interviews with environmental activists from all parts of the world reveal "a striking the-matic pattern: whether the person is from an Asian tropical rain forest, an African savanna, a Latin-American city, a European valley, or a North American farm, they tell a similar story. They have fond memories of a spe-cial childhood place" that they bonded with and can still vividly describe. Yet in all too many cases this landscape of childhood has been altered or even destroyed, and "people relate similar stories about how their special places have changed."[1]

Cultures engage in place-making on a larger scale, whether in establish-ing cemeteries, churches, and public buildings or in deciding to designate an area as a park or a wilderness area. The Navajo (or Diné), for example, have on average at least one sacred place per 26 square miles of their lands. Some are ruins, but most are natural features of the land, including moun-tains, rock outcroppings, bodies of water, canyons, areas where certain plants grow, sites once struck by lightning, and places where a certain echo effect can be heard. The Navajo understand these varied locations as parts of stories that often are linked to tribal rituals.[2] Only a few of these locations have also been embraced by white American culture, however. As Yi-Fu Tuan put it, there are few "enduring places that speak to all humanity," and "most monuments cannot survive the decay of their cultural matrix."[3]

The early work of Yi-Fu Tuan and J. B. Jackson transformed the field of human geography and influenced urban studies, environmental pedagogy, and literary criticism. This work asked what social, aesthetic, and formal qualities made places meaningful, and it reshaped urban planning and revealed values in rural landscapes and folkways that preserve environmental quality, biodiversity, and community. Some scholars in the environmental humanities study how, by sharing a sense of place within a supportive community, we can enhance the engagement of stakeholders and connect what we know—both intuitively and through organized study—with what we care deeply about. What people know about a place connects them to it and provides motivation to protect it. Deborah Tall, who comes from Ireland, has recounted how she forged a connection to the Seneca Lake area in upstate New York. It took years, and included gardening, exploring the area, extensive reading about its history, learning the meanings of Iroquois place names, and many discussions with long-term residents. Knowledge of *where* she was fostered knowledge of *who* she was. People reflect on their own sense of place through memory, share this with others, and build a sense of place that unites communities.[4]

In the abstract, it might seem that at first there were uninhabited or empty spaces, which human beings entered and learned to inhabit, transforming each into a specific place. Many Americans well into the last century imagined their history in essentially these terms,[5] and it has been an attractive idea for many settler cultures, which imagine land as having been largely empty space—a blank on the map—before colonizers arrived. This idea is often embodied in spectacular photographs that depict pristine nature, such as those of Ansel Adams. (See figure 2.1). Yet this a fundamentally mistaken way to imagine space or to conceive of the past. Human beings came late in evolution, and did not wander out of Africa into the rest of the world until quite recently. There were always a great many animals, insects, and plants already in any location that human beings entered. Moreover, there were Neanderthals and other hominids living in areas that *Homo sapiens* invaded about 30,000 years ago. In more recent millennia, human beings began to displace other human beings. Ethnographic studies of contact zones and urban studies reveal how contested and fraught the notion of place can be. So the question of power is also front and center for the environmental humanities. Or rather, two questions: "Who is imagined to belong in this place?" and "Who (or what) has been excluded to make this place?" In Deborah Tall's case, she became intensely aware of the Cayuga and the Seneca, who had been forced out of the Geneva area but whose arrowheads occasionally turned up in her garden.

In the worst instances, place-making was preceded by bloody conquests in which aboriginal peoples were displaced, notably in the Spanish invasion of the Aztec empire. But even when there was little human conflict, place-making meant making accommodation with the people and other life forms already present, learning the peculiarities of a location, and developing a way of life based on its possibilities. This is what some biologists call constructing a "human niche" by reshaping surroundings to enhance human thriving.[6] Whether people recognized it or not, this was always a process of co-determination in which the pre-existing life forms played roles. Bees pollinated crops, birds ate some insects but not others, fish spawned in the streams (or perhaps died out because of a dam on the river), and a complex ecological system adjusted to human presence. People also found that some animals were amenable to domestication, and the proliferation of these domestic animals reduced the possibilities for, or entirely eliminated, other species.

Since at least the middle of the nineteenth century, both fiction and narrative nonfiction have dramatized the environmental complexities of modern societies. The opening pages of Rachel Carson's *Silent Spring* described a scene that at first seemed beautiful and pastoral, but in which fish and birds were being devastated by an unnamed pesticide that was also entering the human food chain.[7] George Perkins Marsh likewise employed a dramatic contrast in the opening to *Man and Nature*. After evoking the productive agriculture of the Roman Empire that undergirded its stability and power, Marsh described the decline in fertility as entire provinces, including several "celebrated for the profusion and variety of their spontaneous and cultivated products and for the wealth and social advancement of their inhabitants," were "deserted by civilized man and surrendered to hopeless desolation."[8] Forests disappeared, rich farmlands washed away, springs dried up, rivers shrank to brooks, harbors silted up, and cities vanished. Marsh acknowledged that this decline was due in part to natural forces, but blamed the destruction on Roman laws, destructive agricultural practices, and lack of good forestry.

On a smaller scale, stark contrasts such as those noted by Marsh and Carson can also be found in the works of Barry Lopez and other present-day writers. But the power of narrative derives from more than powerful contrasts. To awaken environmental awareness involves not only learning to see the historical and human forces that shape external landscape but also cultivating what Lopez calls an "interior landscape" that "responds to the character and subtlety of an exterior landscape" but which also is arranged "according to the thread of one's moral, intellectual and spiritual development." Lopez argues that for aboriginal peoples a story "draws on relationships in the exterior landscape and projects them onto the interior landscape" and "the purpose of storytelling is to achieve harmony between the two landscapes." The value of storytelling in their societies lies in "the power to reorder a state of psychological confusion through contact with the pervasive truth of those relationships we call 'the land.'"[9] From this perspective, modern people lack intimate knowledge of any one place. Their interior landscape is disordered, and they see the world in instrumental terms—as resources to be exploited. For such a historical subject, it is difficult, though not impossible, to find a narrative that can harmonize the interior and exterior landscapes. Much nature writing performs this cultural work. The natural sciences establish laws and relationships in the exterior landscape. No one was a keener observer of such things than Carson or Lopez, but they understood that the challenge for post-industrial people is to develop their interior landscape. As Lopez puts it, "the interior landscape is a metaphorical representation of the exterior landscape" in which "truth reveals itself most fully not in dogma but in the paradox, irony, and contradictions that distinguish compelling narratives." The alternative are "failures of imagination: reductionism in science; fundamentalism in religion; fascism in politics."[10]

Henry David Thoreau was sensitive to the complexity of place-making when he built a temporary house near the edge of Walden Pond. He reflected on the area's earlier inhabitants and literally built remnants of an Irish laborer's shack into his own cabin. He understood that he could coax the local soil to grow beans, but that it would immediately revert to a wild mix of native and introduced weeds the moment he stopped. More than most of his Concord neighbors, Thoreau knew the plants and animals of the region. Yet he also knew that much eluded him. He was fascinated by a wild loon that he chased around the pond in a rowboat one night, trying to guess after each dive where the loon was swimming under the water and always failing to guess correctly. Thoreau understood that he was not living alone in nature and drew upon his cultural heritage, both in his pragmatic living arrangements and in his reading.

In Thoreau's era a majority of Americans and Europeans were still farmers or were only a generation removed from agricultural life. For them, the place of memory beyond their own home was the farm or the village. But in the post-industrial present the majority of people do not work on the land, and their experience of place-making in towns and cities is different. City dwellers often identify with an area no larger than one or two streets, and it takes mental effort to embrace a larger area as a neighborhood.[11] In the next ten years, more than 200 million Chinese farmers will migrate to or be resettled in mega-cities. Such rapid migrations not only pose environmental challenges but also force displaced people to invent a new sense of place.

Some people are nostalgic for a simpler life more deeply embedded in natural rhythms. The environmental humanities increasingly tend toward a different view. In industrialized societies, rural and suburban residents generally have larger "carbon footprints" than city dwellers, because people in the countryside drive considerable distances between workplaces, shops, schools, and a home that is expensive to heat and cool. An urban apartment building with shared walls is more heat efficient than a house, and the greater population density of cities makes mass transit feasible. Nor is place-making in cities necessarily cut off from nature. To the contrary, urban planners now search for ways to make cities "greener." They recognize that a city is not outside the natural world but part of it. Streams once buried in culverts are being opened up, green corridors are being established, green roof gardens are being created, and new buildings are designed to harvest wind and solar energy. Even without such innovations, city residents often have a smaller environmental impact than people ostensibly closer to nature.

Though the image of Thoreau making a small house for himself at Walden Pond might seem the paradigmatic example of taking responsible and imaginative possession of a place, the example is not as simple as a looks. Thoreau lived at the site for a little more than two years. He never owned the land, but he came to know it through intensive and good-humored study. His imaginative domain was the town of Concord, including its woods, fields, and pastures. Such place-making did not require land ownership or the building of anything permanent. People visit the site of Thoreau's house today in part because he taught them to value that kind of place-making. Since Thoreau's day many books have appeared that resemble *Walden* in being accounts of a period of relative solitude in a house on the beach (Henry Beston's *The Outermost House*), the dry Southwest (Edward Abbey's *Desert Solitaire*), or the Wisconsin pine barrens (various works by

Aldo Leopold). Today, Thoreauvian sensibilities are also being brought to bear on the city—see Robert Sullivan's *The Meadowlands* or Novella Carpenter's *Farm City*.

Recent British literature has seen an upsurge of "new nature writing" about place. It has to a considerable degree supplanted travel writing as the most popular nonfiction genre, witnessed for example by the international popular and critical success of Helen MacDonald's memoir *H is for Hawk* (2014). Other writers in the genre include Kathleen Jamie, Mark Cocker, Richard Mabey, Tim Dee, Robert Macfarlane, Alice Oswald, Roger Deakin, Olivia Laing, Sarah Wheeler, William Fiennes, Adam Nicholson, and Paul Farley. Unlike Thoreau, they seek less to capture a spirit of place than to understand the deep entanglements between human history and place. In Britain there can be no sense of wilderness, as the land has been settled for thousands of years. Individual works such as Macfarlane's *The Old Ways: A Journey on Foot* and Oswald's "Dunt: a poem for a dried up river" make millennia-old layers of Britain's paths, hedges, and waterways exquisitely clear. In this sense, British nature writing has less in common with the American tradition than with some literature from Europe and Asia. In addition, Britain industrialized earlier than most other countries, and thus has lived for a longer time with the gap that opened up between county and city.[12] Other nations share this sense of the inseparability of nature and culture. Notably, the environmental history of India, as described in *This Fissured Land*, cannot be told as the story of a contrast between wilderness and culture. In India, people have lived in forests and other natural areas for millennia. Much biodiversity has been preserved, not least because some areas are still held sacred.[13] To put this in larger perspective, in the global North the "green" movement may be energized by the middle class, but in the global South it is "centered more on farmers and fishermen who rely on natural resources for their livelihoods."[13]

Thoreau, neither farmer nor fisherman, revitalized his sense of place and rejected the idea that land was simply a commodity. New translations of *Walden* continue to appear. (It is now available in Estonian and in Farsi.) A project to translate Thoreau's text into twenty-first-century English was launched in 2016 by means of an online crowdfunding platform. Such cosmopolitanism reflects also what the literary theorist Ursula Heise describes as a central ambivalence in the place-based imagination of present-day thinkers, artists, and environmentalists. Heise describes how "the

imbrication of local places, ecologies, and cultural practices" has accelerated in recent decades in such a way that critical awareness of particular
places also necessitates engaging "increasing patterns of global connectivity."[15] The local makes sense only when understood, in a larger perspective, as part of the refutation of the global economic forces that treat land
as a mere commodity. Comprehending the sometimes militant dialects of
place-making and embracing the local does not mean turning one's back on
the rest of the world. Quite the contrary, it leads to embracing other parts of
the world through travel, multilingualism, and new forms of tourism. Even
Thoreau did not always remain in Concord; he also wrote about the Maine
woods and about Cape Cod.

Ecological Tourism

Although most people have the chance to live in and know only a small
number of places in detail, increasing numbers now travel the globe and
know many places. For four centuries after Columbus much of this movement was inseparable from European imperialism, which sent out colonial
administrators, teachers, traders, sailors, and soldiers to control and manage the colonies. In a book titled *Green Imperialism*, Richard Grove explains
that this movement included a quest for edenic landscapes and involved
a series of at first unintended experiments in environmental transformation, particularly in island societies.[16] The small scale of islands made it
possible to see in a generation or two how the introduction of new forms
of agriculture and new species could utterly transform the country. Often
there were disastrous consequences, such as the rapid deterioration of the
environment on the island of St. Helena. As Grove makes clear, instances of
rapid decline encouraged Europeans to learn from indigenous cultures and
to develop a more holistic view of nature, new nomenclatures, and new
practices—notably state ownership of and conservation of forests, which
began in mid-nineteenth-century India. In short, imperialism was not simply conquest. It had reciprocal effects on Europeans, who began to understand history and geography in environmental terms.

In the post-imperial age, many people's visits to new places are so brief
that they gain only a casual impression. Often visitors damage sites, for
example by riding bicycles, motorcycles, dune buggies, or snowmobiles
over fragile ecosystems. (See figure 2.2.) Yet for several decades ecological

sensibilities have been reshaping tourists' interests, and the industry has changed accordingly. Ecotourism has been developed in part to minimize tourists' effects on local environments, and this is an important aspect of its appeal. But in addition, an ecotourist wants to understand a new place from the inside, as its inhabitants do. A traditional guidebook explained the history, art, and architecture of a site. An ecotourist wants information about a region's biodiversity, its food ways, its endangered species, and its efforts to reduce carbon emissions. Some ecotourists also want to take part in local environmental projects.

Tourism is one of the world's largest and fastest-growing businesses, now representing 10 percent of the global economy. By the 1970s it had become an academic subject. In a book titled *The Tourist: A New Theory of the Leisure Class*, Dean MacCannell argued not only that tourists were a largely middle-class group in search of experience and titillation, but also that "the tourist" was a cultural formation. The "first apprehension of modern civilization," he declared, "emerges in the mind of the tourist."[17] For the tourist, a site exists first through its representations and "markers," and these dominate and determine its meaning. The novelist Don DeLillo made a similar observation in his novel *White Noise*, in which a great many tourists admire and photograph an exemplary Pennsylvania barn whose iconic status is so well established that none of them really see the actual barn.[18] MacCannell further argued that "tourist imagery is absolutely plastic," with little concern for accuracy so long as it is appealing.[19] Yet many tourists yearn for "typical" cuisine and "authentic" experiences, which the tourism industry presents as a "reality" behind a facade that fools less discerning visitors. Staged authenticity involves creating a "pseudo back region" that tourists briefly see and then feel that they have penetrated behind the scenes.

MacCannell's analysis raises many questions about the role of the tourism industry in presenting or preserving other cultures, much less protecting endangered environments. John Urry carried the critique of tourism further in *The Tourist Gaze*, which drew on Michel Foucault's concept of "the gaze." Urry did not find the essence of tourism to be the search for authenticity. Rather, he examined how the tourist systematically develops a particular form of the gaze that values objects for their contrast with everyday experience—perhaps because of origin (a "moon rock"), historical associations (the spot where Olaf Palme was murdered), fame (Niagara Falls), typicality (a German beer garden), or blatant inauthenticity (Las Vegas). As part of this tourist gaze, Urry noted "the increased fascination of the developed world with the cultural practices of less developed societies" and how the tourist had become "essentially a 'collector' of places often gazed upon and experienced on the surface."[20] Because the tourist gaze has begun to merge with the way shoppers look at commodities, the experienced tourist would seem ill-prepared to engage environmental problems. Nor was the tourist gaze something new, for it emerged along with imperialism. As James Buzard explained in his book *The Beaten Track,* between 1800 and World War I the basic structures of the tourism industry were established and then ironized over by many writers, notably Henry James and E. M. Forster.[21]

In recent decades the tourism industry has developed a portfolio of products for the growing niche market of ecotourism. These include visits to areas that conserve natural resources, visits that benefit the welfare of local people, and visits that give tourists a sense of place. Though there were isolated examples of such travel much earlier, ecotourism emerged as self-conscious practice in the 1990s, and became highly organized on the supply side. Demand has increased more rapidly that the growth of tourism as a whole, and each year ecotourists spend billions of dollars. What counts as "ecological tourism" operates at several scales. In 2014, one fifth of American travelers were willing to pay a little more for a hotel if it had a good environmental record, and 90 percent participated in hotels' towel-reuse programs. National parks, scenic landscapes, and seascapes have long been popular, but a growing percentage of travelers want to know the histories and the cultural meanings of such sites. Some countries—notably Costa Rica, Switzerland, and Botswana—have made sustainability central to their branding. Ecotourism appeals to people who are not satisfied by lying on a beach, and who instead want a vacation during which they can practice environmental and social responsibility.

Whereas ordinary tourists seek escape and entertainment, ecotourists want to learn about local cultures and their ecosystems. Whereas ordinary tourists often book a hotel on the basis of whether it has a pool, air conditioning, or Wi-Fi, an ecotourist books a hotel built from local materials, perhaps with its own solar power or with composting toilets. Indeed, the appeal of such hotels led the Hilton Hotel chain to set several "green tourism" goals to be achieved by 2014: to reduce energy consumption from direct operations by 20 percent, to reduce carbon emissions by 20 percent, and to reduce the output of waste by 20 percent.[22] As this example suggests, it can be misleading to make a firm distinction between ecotourists and other travelers. At least one study found that ecotourists bring with them patterns of consumption that create acute problems (for example, problems having to do with the management of solid waste).[23] In practice, the siting of ecotourist destinations in desirably remote "natural" areas can reproduce neocolonial relationships among tourists, non-resident or expatriate hotel operators, and local residents.

Though the impulses behind it are commendable, ecotourism runs the risk (identified by MacCannell) of becoming an invasive and largely bogus search for authenticity. Many visitors to Lapland, for example, travel with tourist operators from elsewhere, and profits accrue to outsiders who have only a cursory acquaintance with Lapland. Even with the best intentions, the "authenticity" on offer may be superficial. To remedy this, the Sámi people of Lapland began to award certification to operators who offered "genuine Sámi experiences," who coordinated their activities with locals, and whose tours were based on sustainable practices. The goal was both to improve the tourist experience and to incorporate this tourism within the local culture. The 700 member organizations of the International Ecotourism Society encourage such projects, and the society certifies selected "eco destinations" as offering genuinely sustainable tourism. This is surely an improvement. However, as the sociological work of Erving Goffman suggests, tourists have always longed for "authentic" experience. A successful tour operator gives them the sense that they have moved behind the scenes to a "backstage" area.[24] Yet this penetration behind the facade is often itself a staged event that takes place on a schedule, with clearly prescribed roles for both host and tourist.

At its best, ecotourism includes recycling, energy efficiency, water conservation, respect for other species, learning about another culture, and creating economic opportunities for local communities. For example, in Guatemala a project has organized Mayan women weavers. It certifies traditional production and makes it more visible, which encourages tourists to purchase the weavers' work. In contrast, at some Mexican markets, handicrafts being sold as local products are in fact mass-produced imitations from China. The increase in sales of Guatemalan weavings is certainly an improvement for both residents and tourists.

There is also an alternative "dark" form of ecotourism focused on sites of environmental disasters. Viewing the wreckage caused by a hurricane or the deserted houses of Chernobyl may spring from a commendable wish to bear witness to environmental problems; however, rather than a conscious-raising experience, it can be the tourist equivalent of viewing a horror film. "Dark" ecotourism raises ethical concerns, especially when disaster sites are "stage managed" to achieve certain effects. Disasters can also have secondary effects, including wiping out historical memory of past human-environment relations. For example, Anthony Carrigan has compared how post-tsunami Sri Lanka is depicted by the national Tourist Board and how it is depicted in Chandani Lokuge's novel *Turtle Nest*. Carrigan joins postcolonial and ecocritical concepts in reading the linkage of abuse, sex tourism, and the compound economic and ecological crises that disempower beach communities.[25] His work is a model for how the environmental humanities can sift the power dynamics in developing ecotourism in places scarred with legacies of exploitation and disenfranchisement.

The *Journal of Ecotourism* registers in its pages the tension between planning, management, and profiting from ecotourism and the ethical concerns it creates. In fact, the very idea of ecotourism is a Western invention that implicitly assumes a split between nature and culture and between postindustrial societies and the developing world. From there it is only a step to making a distinction between "shallow" ecotourism that aims primarily at a "feel good" experience for the visitor who briefly drops in and "deep" ecotourism that requires an outsider to make a long-term commitment and to engage more strenuously with the complexities of an unfamiliar culture. The idea of nature itself becomes problematic in this situation. For example, a farmer in Belize or Tanzania will not automatically embrace a European's desire to set aside land as a preserve or protect animals that are locally seen as dangerous competitors, much less appreciate privileged access given to outsiders to the new preserve (notably scientists and ecotourists). The farmers of Bakadadji, a village in Senegal, see their fields invaded by marauding warthogs, which are protected by the new conservation park's game wardens, who are paid by international development funds as part of an ecotourism project.[26] The farmers must scrounge for funds needed to build fences to keep the warthogs out, or else they will see their harvests ruined. In developing countries, ecotourism can seem like a banquet set for someone else to eat.

In wrestling with such problems, the environmental humanities can be crucially important, not only in the design of spaces and the preservation and enhancement of local ecologies, but also in assisting nuanced two-way communication between visitors and their hosts. Ecotourism may also have a valuable secondary effect by raising local awareness of environmental issues and increasing pride in a region's natural heritage. A weaving tradition using locally sourced fibers and natural dyes may revive as a result of ecotourism. A community that traditionally hunted whales might limit hunting and supplement its livelihood by taking tourists out to see the whales. With such goals, the United Nations has created a program to help poor countries develop ecological tourism as a route to sustainable growth.[27] For example, the economy of Belize has been stimulated by ecotourism, which represents 15 percent of its GDP. Ecotourism can lead to measurable environmental and economic improvement. The environmental humanities can make a constructive contribution by articulating the environmental history of an area—a history preserved in its songs, literature, legends, landmarks, art, and agricultural practices. With such an understanding, tourism acquires new meanings and destinations. The European Union, recognizing the potential value of such work, has created a European Ecotourism Knowledge Network.

Ecotourism can have negative consequences, however. For example, in the late 1960s and the 1970s, when ecotourism was just emerging out of the counterculture, the Himalayan Buddhist kingdom of Nepal seemed to "dharma bums" on the international road to be an enchanting counterpoint to Western commercial culture. Thousands of tourists flocked there. The local government responded by creating a national park and developing tourist destinations in the mountains and in the low-lying jungle. Yet even as Nepal strove to present itself as a paragon of biodiversity, "the valley of Kathmandu—which had seemed like a fairy tale out of the *Thousand and One Nights* as late as the 1970s" was increasingly "suffocated by exhaust, smog, noise, and garbage."[28]

Indeed, "ecological tourism" is a somewhat self-contradictory concept. To the extent that travelers arrive in airplanes or automobiles, they add to global pollution levels. Large numbers of tourists make demands on local water and food resources, and they often drive up property values by purchasing vacation homes. Tourism also increases traffic and can lead to road construction that is costly to local government and that adds to air pollution. Setting up a nature reserve to protect an endangered species can attract too many visitors. Demands for water and sanitation can harm the local ecological system, and foot traffic along trails can damage tree roots, trample plants, and scare away the very species being protected.

Problems similar to those caused by ecotourists trekking in Nepal emerged at the Monteverde Cloud Forest Reserve in northern Costa Rica (once famous as the only site where one could see the golden toad, which now appears to be extinct). The reserve also protects other species, including more than 2,000 flowering plants and more than 100 mammals. The Galapagos Islands, an Ecuadorian national park and a UNESCO World Heritage Site, have experienced pressure from tourism and a growing human population that has been documented by Cecilia Alvear and other native journalists.[29] In short, ecotourism, even with the best intentions, can love endangered species to death and shift costs to local communities.

The small country of Bhutan has taken a completely different approach to tourism than its neighbor Nepal. Bhutan permits only a few tourists to enter the country per year. In 1969 it nationalized its extensive forests, and after 1980 it pursued a policy of reforestation. It also created a "green belt" of uninhabited lands along its border with India. It managed to avoid the rapid population growth that Nepal experienced. Instead of encouraging tourism as a source of income, Bhutan avoided extensive contact with other cultures; it even expelled a growing minority of Nepalese immigrants.[30] To some observers Bhutan exemplifies a society with a strong sense of place that resists globalization, but to others it appears xenophobic. Its example certainly shows that ecotourism is not universally embraced, and that quotas or limits may become necessary to avoid degrading endangered environments.

Precisely because ecotourism is a complex and culturally ambivalent practice, it benefits when historians ensure that it is based on accurate information, when anthropologists mediate the dialogue between tour operators and local cultures, and when short-term profits are not permitted to drive the enterprise. The value of ecotourism is not limited to sites visited, for travelers return home with sharpened sensibilities. After seeing exemplary practices elsewhere, one may wish to improve practices in one's home country. An American who has seen the extensive use of bicycles in Denmark may seek to introduce cycling paths in his or her home town. An Indian who has seen the success of solar power in Germany may want to see it adopted in New Delhi. An Australian who came to prefer a high-fiber diet while visiting South America may plant a home garden with such a diet in mind. A Nigerian who has witnessed careful recycling in the Netherlands may be inspired to introduce similar practices in Lagos. Clever reuse of vegetable oil tins in a Nigerian street kitchen may influence industrial design in Eindhoven. Through many such encounters, ecotourism can stimulate new practices and foster useful projects on the traveler's home terrain. Ideally, it can foster not globalized homogeneity but the sharing of best practices that are adapted to each local economy and each local ecology.

The New Wilds

The lives of large wild carnivores cross political boundaries. Wolves have dispersed across the US-Canada border and from the Italian Apennines into France; jaguars enter Arizona from Mexico.[31] The longest ongoing study of wild animals concerns the wolves of Isle Royale, a 210-square-mile (544-km^2) island in Lake Superior. Isle Royale is a designated federal Wilderness Area and has been preserved since 1940. After it became a natural laboratory, wolves crossed the frozen lake from Canada in the winter of 1949, as they apparently had been doing for centuries. Wolves are the island's only large carnivores; moose are its only large ungulates and the wolves' main prey. In 1958, Durward Allen and L. David Mech launched a study of changes in the populations of wolves and moose; that study has continued, though its future is uncertain. Global warming has raised winter temperatures so that Lake Superior freezes less often, and thus the island is isolated for longer periods than it once was.

The wolves of Isle Royale have become a model for wildlife managers who wish to protect large carnivores, which often are seen as important to the preservation of wild ecosystems. For 60 years, dozens of researchers and volunteers have counted the wolves from the air and have scoured the island collecting bones. After a parvovirus devastated the wolf population in the 1980s, David Mech and others pioneered the use of radio collars to monitor the population. Collaring and tracking wolves raised howls from wilderness purists, but the practice probably reduced the incidence of violent interactions between farmers, ranchers, and wolves across western North America. Research-based advocacy has helped rehabilitate wolves' reputation. David Mech and Rolf Peterson are known internationally for their popular writing and their photographs as well as for their scientific publications. Their work helped to overturn mistaken assumptions about a "balance of nature" between wild predators and prey and to transform people's perceptions of wolves, which have gone from being one of the most reviled species to being a symbol of wildness.

Though wolves may have become more popular, the wolves of Isle Royale are threatened with extinction. The population declined from more than 30 individuals in three packs in the 1950s to seven in a single pack by 2011, and the 2015–2016 report listed the probable population as two.[32] By raising questions about the future of wolves on Isle Royale, Mech and Peterson set in motion a cultural change in humans' relationships to wild areas.[33] Questions of ethics, politics, culture, and policy require collaboration between public stakeholders and scholars across many fields. Should the human guardians of wilderness transport new wolves to the island? Should the moose be allowed to proliferate until they overgraze the island, destroying their own habitat? What degree of intervention is called for to preserve wildness? And just what *is* wild about Isle Royale, with its closely studied collared wolves, its aerial counts, and its seasonal stream of human visitors who visit the island for recreation?

Wild animals now make up less than 5 percent of the vertebrate biomass, while humans and domesticated animals account for the rest. Yet conservation of wildlands remains one of the most emotionally charged (and politically weighted) discussions about the environment. In the 1960s, preserving wilderness areas seemed to be an ideal way to protect natural diversity. However, environmental historians successfully challenged the concept of wilderness in the 1990s, and the result was a slow change in how public lands and natural areas are managed. Management now more often involves local stakeholders whose presence and long-term use of natural resources had been overlooked or excluded, at times to the detriment of biodiversity.

More recent research by geographers and anthropologists feeds back into the management of natural resources and lands. The measurable effects of climate change and economic globalization present new challenges to the protection of places once designated as "wilderness," and managers and scientists have turned to new categories, such as novel and hybrid ecosystems that have some interaction with humanity but remain beyond human control. The environmental journalist Emma Marris has argued that the earth is best seen as a "rambunctious garden," with humans as caretakers rather than masters.[34] Nature reserves populated with a mix of native, introduced, and hybrid species manifest unpredictable qualities and forms of life; the anthropologist Eben Kirksey describes such places as "emergent ecologies."[35] The wildness of areas lightly inhabited by humans, who hunt and may themselves become prey, persists. We need to attend "to our relations with those beings that exist in some way beyond the human," writes the anthropologist Eduardo Kohn in his multispecies ethnography of the Ecuadorian Amazon. Considering the shaman who speaks of himself as a man-jaguar, *runa puma*, "forces us to question our tidy answers about the human."[36] If we pry open the categories "human" and "wild," we find that much Western thinking about wilderness reflects the alpine forests treasured by a privileged minority of Euro-Americans, mostly affluent, white, and male. This idealized alpine landscape, projected on to North American landscapes and reborn as "national parks," persisted for two centuries. Perhaps it even delayed recognizing ecological value and providing legal protection to wetlands, coastal areas, marine sanctuaries, and grasslands.

Biologists and environmental historians have learned to see wild areas such as Isle Royale as far from "pristine," though in no less need of caring human attention. The ur-wilds of a generation ago—Amazonia, the American West—are now known to have been long inhabited, their soils freighted with the bones and charcoal of past human settlement and agroforestry, their vegetation and wildlife matrices reflecting centuries of change via nonhuman *and* human influences. Yet wildness of a hybrid, non-pristine sort is also spreading locally even as the human population expands globally. Rural depopulation in Europe and North America in the wake of urbanization and the industrialization of agriculture has led to reforestation and to the return of large mammals (including some never-domesticated species such as elk, moose, beaver, and eagles) to places from which they had long been absent, including upstate New York, the Carolinas, Norway, and the Baltic states.[37]

Should governments "cultivate" wild places and encourage benign neglect of demilitarized and deindustrialized zones? The government of the Netherlands, for example, has undertaken a project to introduce wild horses and other large herbivores to a 6,000-acre strip of polder (new land produced by draining the sea) called the Oostvaardersplassen. When plans for industrial development faltered, biologists persuaded the government to permit them to "restore" a Paleolithic landscape with imported archaic cattle breeds, Konik horses from Poland, and Scottish red deer.[38] Foxes, egrets, and geese found their own way, and flocks of waterfowl now breed in what the filmmaker Mark Verkerk labeled "de nieuwe wildernis."[39] What are the social, cultural, and psychological benefits of de-centering human civilization by making way for new wilds? Are we ready to move forward to a new concept of wildness, one that is post-wilderness if not post-nature?

The idea of wilderness, or wild lands that have never been farmed or logged and remain unmarred by more permanent human marks, is both very old and hotly contested. Wilderness helped to define civilization as its polar opposite during the modern era in Western Europe and the Americas, and it functioned as a key category of imagining modernity and of expanding empires.[40] Particularly in Christian settler cultures from North America to Oceania, wilderness was a pivotal concept in the political and moral vocabulary. Imagined as an unpeopled paradise, it became the locus of nationalism, a proving ground for "strenuous masculine living," and a zone of conquest.[41] As William Cronon pointed out in his influential essay "The Trouble with Wilderness," "The myth of the wilderness as 'virgin' uninhabited land had always been especially cruel when seen from the perspective of the Indians who had once called that land home."[42] In the United States, wilderness received a clear federal definition in the Wilderness Act of 1964 (Public Law 88–577) as "an area where the earth and its community of life are untrammeled by man, where man himself is a visitor who does not remain."[43] From North America, this legislative, territorial concept of wilderness traveled as an object of science and policy through international institutions and national agencies across the Americas, Oceania, and (via American and European organizations) Africa and Asia.

When the concept of wilderness was exported, it overwrote and even criminalized existing traditions of land use, banning hunting and foraging. Management approaches developed for a "New World" assumed to be without human history were awkwardly adapted in Africa and southeast Asia. Forest-dwelling peoples with intimate knowledge of plants and animals and with ongoing practices of hunting and foraging have faced a wave of exclusion that has resembled the colonial appropriation of common forest resources for export.[44] Around the world, as in North America, wilderness parks and conservation areas became sites of conflict, protected if not established at the point of a gun.[45] Often the interests of big-game hunters and international logging companies have been a stronger force in shaping and policing nature reserves in Africa than the rural communities who are blamed for continued environmental degradation.[46] In many countries, the language of national conservation and parks was borrowed directly from the 1964 Wilderness Act, and afterwards indigenous groups were redefined as outsiders suing for visitation rights to former home grounds. Fortunately, many countries (including Ecuador, Brazil, Kenya, and Tanzania) have moved toward community-based management based on a new wilderness paradigm rooted in historical insights that see humanity as part of nature.

Contests over inhabited wilderness also illustrate how wilderness and wildness are registered differently across cultures. "Wilderness Babel," a 2013 digital exhibition organized by Marcus Hall, makes evident the linguistic diversity of concepts in languages other than English that overlap with ideas expressed in the Anglo-American term "wilderness."[47] As the exhibition's name suggests, Hall's team did not find easy consensus and few direct translations, because many cultures do not distinguish between land types based on human absence.[48] In some languages, another word seems to function much like "wilderness" in public discussions of conservation. Like "wilderness," the Swedish word *vildmark* has frightening connotations and also conjures up a romantic attachment to a rural past imagined to be simpler and more rugged.[49] *Vildmark* names landscapes that are considered "wastes" because they are unsuitable for agriculture but also promises extraordinary, even extreme experiences to tourists. It is the locus of timber rafting, hunting, camping, and the "adventure" of life at the edge of an old resource extraction economy—or perhaps of a vacation spent in a summer house up north.

The languages of sedentary agricultural communities, in contrast, do not distinguish between land uses on the basis of whether humans are present or absent. In Estonian, peasants' reliance on the land is matched by a range of specific terms for land types that are more nuanced than imported concepts that translate the concept of wilderness. Estonia's biosemioticians may hear spring birdsong in the forest as the country's unofficial national anthem, and may take that avian-human communication as a sign of wild culture, not as a wilderness defined by human absence.[50]

Similarly, in Japan mountains and forests were traditionally honored for religious reasons along with wild animals, including those now extinct, like the wolf.[51] The country's Ministry of Environment, in turn, uses the term *gensei-hogo* for the five national "wilderness areas."[52] For the Ministry, the semantic distinction names a higher level of protection. At the same time, visitors to Fujisan (Mount Fuji) might be seeking a spiritual experience, whether or not they reach the summit. Today's Fujisan tourism resembles the secular mountain-worship of European Romantics at Mont Blanc or Mount Snowdon, or Thoreau's experience on Mount Katahdin. Although much of the summit lies within a national park, Fujisan is not wilderness. Indeed, in designating it a World Heritage Site in 2013, UNESCO noted that the mountain has been home to religious shrines since the twelfth century and that for about 1,000 years it has inspired poems, paintings, and other artistic representations. Even though land management in Fuji-Hakone-Izu National Park may have been influenced by North American practices and Western ideas of wilderness, the mix of reverence, awareness of long human inhabitation, and national symbolism makes "heritage" a more apt category than "wilderness."

Despite the inconsistencies of the Western wilderness ideal, the romantic cult of wild places retains a powerful appeal for non-government organizations, artists, and political movements. The Brazilian photographer Sebastião Salgado's exhibition "Genesis," which presents a view of the earth as still largely untamed and resplendent, is one of the most widely viewed single collection of photographs to date. What accounts for the undiminished allure of wildness? How might the environmental humanities learn from Salgado and other artists?

Salgado has founded a center for restoration (Instituto Terra) and a national park in his native state of Minas Gerais, which, as a 2015 mining disaster revealed, is both teeming with diverse life and under threat. Salgado emerged as an acute observer of social inequality and injustice with two critically acclaimed books of black-and-white photographs, *Workers* (1993) and *Migrations* (2000). By his own account, the experience of documenting migrations in the 1990s and the incredible brutality and violence that hounded refugees from armed conflicts and natural disasters led him to lose faith in the future. With his wife, Lelia Deluiz Wanick, he began a long-term ecological restoration project on their family's property in the Vale do Rio Doce of Brazil. They planted more than 300 species of native trees. Reforestation brought a resurgence of wildlife—not only beetles and butterflies, but also alligators and fish.

Salgado's next major photographic project resulted from this experience of local rewilding and ecological restoration. "My goal," he later recalled, "was not to go where man had never before set foot, although untamed nature is usually to be found in pretty inaccessible places. I simply wanted to show nature at its best wherever I found it."[53] The resulting images had the dual purpose of inspiring hope in the possibility of a more balanced coexistence of humanity and nature[54] and warning of planetary endangerment by modern ways—including endangered lifeways of remote peoples such as the Stone Korowai of West Papua and the peoples of the Upper Xingu Basin in central Brazil. The "Genesis" exhibition toured worldwide. Its popularity testifies to the allure of romantic wild nature and the intensity of the tourist gaze it lavishes on human subjects—who often look back at the viewer.

In part, the success of "Genesis" stems from its confirmation of widely marketed conventions concerning so-called primitive peoples and the majesty of wide open spaces in Africa, in Amazonia, and at the poles. The landscape photographs recall the epic style of Ansel Adams, and the individual portraits recall traditional Western iconography. Naked human figures are captured in neoclassical poses, and their arrangement in the book based on the exhibition often evokes a specifically Judeo-Christian conception of creation. For example, a young girl preparing for the Amuricumã ceremony appears in a photographic diptych opposite a male athlete painting himself for Kuarup wrestling; the diptych unfolds to reveal a series of images of female-centered and male-centered festivities in a village in southern Amazonia.[55] In *Genesis*, wild nature is thickly populated but reflects more static, Edenic conventions and a kind of reverential neo-primitivism. What of the anthropological, ecological dynamism of wild places?

Salgado's photographs of the Korowai of Papua New Guinea court further controversy. The Korowai are notorious figures of the neo-primitivist Western anthropological imagination. For decades they were cited as an example of a "Stone Age people" and eroticized for their semi-nakedness and the males' elaborate penis sheaths. None of Salgado's photographs included in the book *Genesis* depict Korowai in modern Western clothing, though the wearing of such clothing has spread as a result of the influence of Christian missionaries. In an analysis of stereotyping between tourists and the Korowai, the anthropologist Rupert Stasch has argued for a radical symmetry in which exoticism and orientalism characterize Western portrayals of the Korowai and other "primitive" peoples, but the Korowai in turn have a range of stereotypes for Western tourists, who are variously feared, hated, and wooed as a material resource.[56] Ecomedia critics, anthropologists, and artists are engaged in a dialogue about the place of humans in wilds and about the aesthetic principles and ethical ideals that are often read from and projected onto landscapes and "primitive" peoples by Westerners. The humanities, by developing a sense of place, improving ecotourism, and redefining wilderness, can help to discredit stereotypes and stimulate further dialogue between cultures.

Despite its reliance on old landscape conventions as visual stereotypes, Salgado's presentation of landscape, wildlife, and human subjects works toward a radically egalitarian aesthetics and a vital perception of earthly wildness. One sequence of four photographs epitomizes Salgado's egalitarian lens at work. A 2009 portrait of a Zo'e woman holding a child at the edge of a shallow wetland in the northern Brazilian state of Pará is followed by a 2011 image of a great egret in Brazil's Pantanal region. A generous depth of field renders background foliage patterns as crisply as foreground subjects, so the eye must work to pick through the image and discern habitat and organisms. Seeing thus becomes an ecological act. The next two images present a visual echo between reptilian scales and fissured, incised rock outcropping: one hind leg and the partially submerged tail of a yacara caiman are reflected in a diagonal of water that spills with the page's turn into a river below the incised, reptilian-scale rock face of La Cueva de Auyan in Uruyen, Venezuela. Environmental humanities approaches to rewilding and wildlife can learn from Salgado's visual technique, which provokes awe, sympathy, and a kind of ecological leveling of perception that does not, ultimately, place the human figure above other creaturely forms.

Genesis suggests that it is possible to tell a story of wilderness and wilds that is not centered in a Euro-American genealogy but dispersed in and emerging from millennia of wild inhabitation across the earth. This dispersed planetary wildness was apprehended by the poet Gary Snyder in a book titled *The Practice of the Wild*: "[W]ildness is not limited to the 2 percent formal wilderness areas. Shifting scales, it is everywhere: ineradicable populations of fungi, moss, mold, yeasts, and such that surround and inhabit us."[57] Sacred groves, mountain peaks, and desert oases have been preserved and defended for millennia as windows into the wild. Tibetan circumambulation of sacred peaks, Aboriginal mapping of watering holes and trails, aristocratic hunting parks in China, Hindu temple forests: these pockets of the wild preceded the romantic fascination with unpopulated wilderness that was embedded in Western political, legal, and scientific discourse. Older traditions imagined humanity as part of nature, and they have resisted the absolutism of wilderness thinking.

While dominant religious institutions have sanctioned human expansion by defining the wilds as the devil's territory—to be subdued and conquered— counter traditions of valuing wildness persist in many religions. Ancient groves are preserved around Shinto and Buddhist shrines in Japan and Hindu temples in India; summits are revered in Mongolia; desert sanctuaries in the Middle East and Northern Africa long used by religious minorities survive at a safe distance from Mecca and Jerusalem. One task for the environmental humanities is to investigate the plural, multilingual, and often spiritual bases for renewing wild margins around the world.

Perhaps the newest and most threatened wilds are underwater in the world's warming oceans. The spring of 2016 witnessed the most extreme coral bleaching event yet recorded, with more than 90 percent of Australia's Great Barrier Reef affected. The damage, which may be permanent, takes a cultural as well as a biological toll. In *The Reef: A Passionate History*, the historian Iain McCalman describes how the Great Barrier Reef exists as the world's greatest marine living environment in large part as a testament of the human heart and mind. Plans to mine the reef were countered by passionate resistance that came not from a radical minority or a privileged elite but from a vast coalition of Indigenous people, scientists, artists, and castaways.[58] Such unlikely and numerous allies are necessary to defend the wilds. Neoliberal governments champion a view of oceans as a blank slate and seek to transform the wildest regions of marine biodiversity into human property that generates a revenue stream. But dead coral reefs are a dead-end investment.

Figure 3.1
The Hellisheiði Geothermal Energy Plant, Iceland's largest power plant and the world's second-largest geothermal power station. Located on a volcano, it extracts energy through fifty wells that are between 1,000 and 2,000 meters deep. UN Photo/ Mark Garten

3 Energy, Consumption, and Sustainable Cities

Energy long seemed inseparable from progress and economic growth. Since about 1970, however, the understanding of energy has changed fundamentally, and it now is understood to be inseparable from many environmental problems and their solutions. Energy use is also a central aspect of consumption and of how and where people live. Though environmentalism has long been associated with the countryside, half of the world's people now live in cities. By 2030, according to United Nations estimates, 60 percent of the population will be urban and 95 percent of urban growth will be in developing countries.[1] How can cities, which occupy only 3 percent of the earth's land area, become "greener," with more efficient and sustainable energy use?

Energy

Just before the emergence of the environmental humanities, the cultural meanings associated with energy began to shift dramatically. From the eighteenth century until the 1960s, progress was associated with increasing mastery of energy. The rise of Europe and its colonies to become global empires was based on energy systems, particularly steam power and electricity. The anthropologist Leslie White declared in 1949 that "culture evolves as the amount of energy harnessed per capita per year is increased." It then seemed incontrovertible that "the degree of civilization of any epoch, people or group of peoples, is measured by ability to utilize energy for human advancement or needs."[2] That kind of thinking created a hierarchy that mapped historical development from hunters and gatherers (who had only their muscle power) to farmers (with their domesticated animals) to societies that controlled windmills and watermills, and so on up to the apex of the system: the few countries with atomic power. A similar conception of social evolution had long been expressed at world's fairs, which gave enormous prestige to science and engineering. The development and exploitation of oil, coal, natural gas, and hydropower seemed crucial to prosperity. Rising levels of energy consumption became indicators of progress.

This view of energy's role in history began to change during the 1960s. People started to realize that energy production unavoidably caused waste and pollution. Strip mining for coal destroyed enormous tracts of land, and burning that coal produced smoke pollution and acid rain. Likewise, drilling for oil was hardly a clean, surgical procedure. It spewed gushers of crude oil over the landscape, shot tons of gas into the atmosphere, and demanded swathes of land for new highways, pipelines, and other infrastructure. The discovery of oil on the northern coast of Alaska provoked a battle between environmentalists and oil interests that also involved Native Americans. The latter could profit from leasing their tribal property and selling their mineral rights, but the land might be polluted beyond recognition. The oil interests won this argument, in part because it occurred in the midst of the oil shortages of the 1970s. This "energy crisis" began during the first Nixon administration. Some argue that it was engineered by the oil corporations themselves; others contend that it was caused by demand outpacing supply in a process that began before 1970.[3] The shortages became acute in 1973, when members the Organization of the Petroleum Exporting Countries (OPEC) began to limit their output. Demand quickly outraced supply, and prices skyrocketed. The high prices of oil and gasoline foregrounded the importance of energy in advanced economies and contributed to a recession in Europe and the United States.

Some advocated nuclear power as the solution. In the 1950s atomic power plants were expected to produce energy that would be "too cheap to meter," but construction, maintenance, and security proved more expensive than had been estimated. France converted to nuclear energy, but in many countries there was resistance from environmental groups. The resistance became overwhelming after the accidents at Three Mile Island in the United States and at Chernobyl in the Soviet Union. Yet even without such accidents, atomic energy had a serious long-term drawback: waste. The American nuclear power industry and the American armed forces produced more than 100 million gallons of radioactive wastes without paying into a fund to deal with them. This waste is stored at more than 150 sites in 40 states.[4] Just a few gallons of it, evenly distributed, could kill much of the American population. There are also solid wastes. At the Hanford Site in the state of Washington alone there are "25 tons of solid plutonium" that "must be kept under constant armed guard" because even a few kilograms of it would suffice to build a nuclear bomb. At Hanford, 11,000 workers are attempting to clean up the 586-square-mile site, at a cost of $2 billion a year. The worst dangers can be removed, but it "is essentially uncleanable and will remain hazardous" for tens of thousands of years.[5] In short, not only is nuclear material dangerous during use; when the costs of storing and guarding nuclear waste are added to the utility bill, nuclear power is a short-term solution with extremely high long-term costs for future generations.

It is little wonder that by the end of the 1970s many rejected the idea that high energy use was a sign of progress. Instead, they emphasized energy efficiency. It proved possible to extract more energy from each ton of coal or barrel of oil, and it was also possible to design machines so they would use less energy to achieve the same results. This made industrial countries less dependent on oil imports. Moreover, in subsequent decades it gradually became clear that energy production also was an important driver of global warming. The burning of petroleum in automobiles, ships, and tractors and the burning of coal in power plants produced the majority of the CO_2 and other pollutants that have caused the "greenhouse effect," pushing up the world's average temperatures.

Energy consumption is not driven by needs, and it varies considerably from one culture to another. The per capita energy use of the United States is almost twice that of Europe, for example, and one of the tasks of the environmental humanities is to understand the cultural drivers of such variations. Historically, there is no clear correlation between high energy use and a high standard of living. When Germany was divided between East and West, the East Germans used more energy per capita than the West Germans but had a lower standard of living and a higher level of pollution. There is considerable variation in the United States; for example, Massachusetts uses only half as much energy per capita as Texas. Much energy saving has occurred in all parts of the country, however. Whereas many automobiles of the 1950s got fewer than 10 miles per gallon of gasoline, automobiles in 2015 *averaged* well over 20 mpg. In 2017 the most efficient cars got better than 35 mpg, and the best hybrid gas-electric cars were better yet. Likewise, refrigerators, washing machines, and other home appliances improved incrementally to use far less energy.⁶ Engineers improved these products, but convincing consumers to pursue energy efficiency demands skills in communication and an understanding of the place of energy in each culture.

The environmental humanities reject the idea of a cultural hierarchy in which the "primitive" cultures are evolving toward the "advanced level" of high-energy Western cultures. Instead, it now seems evident that high energy consumption is environmentally destructive when the energy comes from fossil fuels. If any culture is to be assigned a higher status, it might be the culture with the smallest "carbon footprint," whether because it has a less demanding pattern of consumption or greater efficiency or both. High energy consumption is now understood not as an achievement but as a problem. It should not be put at the center of a narrative of progress. Nor can energy be understood simply as a scientific problem or an engineering challenge. In fact, by 2017 technologies of energy production already existed to solve the problem of global warming. Not only were solar and wind power competitive economically, but some sites, notably in Iceland, drew heat directly from beneath the earth. (See figure 3.1.) Furthermore, in most economies, total energy consumption could be cut in half without reducing the quality of life. However, energy has become deeply intertwined with the structure of everyday existence. Reducing energy use is now primarily a cultural and political problem, not a technical one. Navigating a transition to post-fossil-fuel energy sources while simultaneously reducing consumption in high-use countries (such as the United States, Australia, and Canada) and increasing per capita energy access across Africa and South Asia will require all the cultural skills of humanists working alongside social scientists, engineers, and NGOs in the field. And some of the primary questions are ethical and political rather than purely technical in nature: By how much will North Americans and Europeans reduce their per capita consumption to offset their historical contributions to climate change? How much more energy for household cooking and lighting do children and families in Ulaanbaatar and Mumbai need to live healthy, productive lives?

The popular phrase "sustainable growth" suggests that humanity can pursue growth much as before so long as it is tweaked a bit to use less energy. It suggests that countries can continue to expand their economies, and that families can continue to grow, while somehow reducing their carbon footprint. But in practice this has proved difficult. The Japanese environmentalist Yoichi Kaya developed a formula, commonly called the Kaya Identity, that explains why. Kaya calculated the total CO_2 emissions for a country by considering four factors: change in population, change in per capita gross domestic product, change in energy intensity, and changes in carbon intensity. According to Kaya's formula, it will not be enough to improve energy intensity (for example, by making machines and processes more energy efficient) and to reduce carbon intensity (for example, by adopting windmills) if the population grows or if the population increases its per capita energy use.

Population growth is not discussed as much today as it was in the 1970s, yet it must not be overlooked. Until recently, China actively tried to limit its population with a one-child-per-couple policy that it has now abandoned because it has resulted in a skewed population profile, with more people retiring than entering the workforce. But an even worse policy is that of India, whose population surged from 350 million to more than 1.2 billion between 1947 and 2012. During the same period, the population of the United States went from approximately 150 million to more than 300 million. For the world as a whole, population growth during the twentieth century has vastly outweighed improvements in its per capita carbon emissions. Consider the last two decades of the twentieth century, during which the world's population increased by 1.6 percent per year. That may not sound like much, but it resulted in an increase of 37 percent during those twenty years. If all those additional people had consumed at the same rate as in 1980, then energy efficiency would have had to improve by 1.6 percent a year, or 37 percent over two decades, to keep pollution at the same level. That did not happen.

Between 1980 and 1999, as the world's population soared, so did per capita gross domestic product, the second factor in the Kaya Identity. Worldwide consumption increased by an annual average of 1.28 percent. There were more people every year, and they consumed more. In China and India, many bought refrigerators and other appliances for the first time. In all parts of the world, more cars were put on the roads. Even in an efficient economy such as Germany, the manufacturing of cars "generated about 29 tons of waste for every ton of car," and "making a car emitted as much air pollution as did driving a car for 10 years."[7] Multiplying the first two factors of the Kaya Identity together ($1.6 \times 1.28 = 2.05$ percent per year) suggests that overall energy demand rose 50 percent over that twenty- year period.

Fortunately, the third and fourth factors somewhat mitigated the increase. The third factor, energy intensity, improved at an annual rate of 1.12 percent. Coal was burned more efficiently, cars and trucks got better mileage, and more efficient industrial machines and home appliances used less electricity than before. Between 1975 and 1985, California raised its building standards and reduced the energy used in a new home by 50 percent per square foot. These measures saved the equivalent of 2.5 gigawatts of electricity every year. Since that time, new Californian homes have been built to higher standards, and many other states have followed suit.[8] By 2008 the American economy used only half as much energy to produce one dollar of GDP as it had in 1970.[9] One study covering the years 1949–2006 concluded that every kilowatt-hour of electricity used in computers, telephones, and other communication devices saved between 6 and 14 kilowatt-hours elsewhere in the economy. Sending email messages and attachments uses far less energy than sending letters and packages. The delivery of physical mail has also been improved by the use of computer programs. For example, one parcel post service developed a program that optimized its routes and saved 3 million gallons of gasoline per year.[10] In short, higher energy intensity has reduced the environmental impacts of population growth and higher consumption.

The Kaya Identity also recognizes that not all forms of energy generation contribute the same amount of CO_2. Its fourth factor is carbon intensity, or how much CO_2 is released by various forms of energy. If more people use mass transit, it will reduce carbon intensity, for example. Countries with large hydroelectric resources, such as Canada and Norway, score better on carbon intensity than Australia, which relies heavily on coal. Indeed, Australia made no improvement on this factor between 1980 and 2000. Adoption of nuclear power also reduces CO_2 emissions, and by 1998 there were 437 nuclear power plants distributed among 29 countries. They lowered carbon intensity, but in almost all cases these plants were built and operated with government subsidies and tax breaks. When Britain privatized its electricity supply industry in the 1980s, the government was unable to sell any of the nuclear plants, because they could not be run at a profit in the private sector. (Wind and solar power offer promising alternatives for lowering carbon intensity and will be discussed below.)

Table 3.1
Annual percent change in Kaya Identity for selected regions and large countries, 1980–1999.

	Population	GDP	Energy intensity	Carbon intensity	CO_2 emissions	20-year growth in CO_2 emissions (1980 = 100)
Eastern Europe	0.44	−1.91	−0.14	−0.61	−2.21	−45.0
OECD Europe	0.53	1.73	−1.00	−1.06	0.18	103.7
United States	0.96	2.15	−1.64	−0.21	1.23	127.7
World	**1.60**	**1.28**	**−1.12**	**−0.45**	**1.30**	**129.5**
Japan	0.41	2.62	−0.57	−0.96	1.47	133.9
Africa	2.54	−0.58	0.82	−0.01	2.77	172.7
Australia	1.36	1.98	−0.37	0.00	2.98	179.9
Brazil	1.61	0.76	1.83	−0.80	3.43	196.3
China	1.37	8.54	−5.22	−0.26	4.00	219.1
Middle East	2.98	0.04	2.45	−1.14	4.34	233.0
India	2.04	3.54	0.27	0.03	5.97	318.9
East Asia	1.78	5.00	0.92	−0.70	7.10	394.3

Source: Jefferson W. Tester, Elisabeth M. Drake, Michael J. Driscoll, Michael W. Golay, and William A. Peters, *Sustainable Energy: Choosing Among Options* (MIT Press, 2005), 26

In summary, the Kaya Identity includes four factors that must be considered together when calculating whether sustainability is being achieved: change in population, change in GDP per capita, change in energy intensity, and change in carbon intensity. When these four factors are considered together, it is evident that different regions of the world performed quite differently in the final two decades of the twentieth century.

East Asia and India performed the worst. As their economies and populations grew rapidly, their total carbon dioxide emissions more than tripled. China's one-child-per-family policy held population growth down and improved energy intensity, but China's economy grew on average more than 8 percent per year, with the overall result that CO_2 emissions increased by more than 200 percent. By comparison, the United States increased its carbon dioxide emissions much less, but it must be kept in mind that in 1980 the US was by far the world's largest producer of CO_2. The Kaya Identity is a measure of the percentage of change. In absolute terms, the US was still the world's largest contributor to global warming in 2000, although today China holds that dubious distinction.

Eastern Europe was the only region to reduce its carbon emissions between 1980 and 1999, and it did so by an impressive 2.21 percent per year. However, this improvement was the direct result of a falling standard of living, as Eastern European countries' economies shrank and life expectancies fell. There was almost no population growth. In addition, during the collapse of the Soviet system, inefficient factories closed, and energy-intensive technologies that had been developed in the West were easily available for the first time. In short, Eastern Europe's good statistical result was a poor result for its people, whose standard of living fell. A plague might produce a similar result.

The OECD countries in Europe offer a better example of what long-term sustainable development might look like. The standard of living rose at an annualized rate of 1.74 percent, and the population grew 0.68 percent a year, but these increases were almost entirely compensated for by improved energy intensity and reduced carbon intensity. Overall, carbon emissions grew by only 0.18 percent a year. Western Europe as a whole came tantalizingly close to zero growth in CO_2 emissions. Several economies, notably Britain, Germany, and France, essentially achieved that goal. The differences between Eastern and Western Europe illustrate John Deutch's observation that "the Kaya relation helpfully identifies the restricted range of choices that a country faces in achieving carbon reduction targets and the significant differences in choices likely to be made in different countries."[11]

How and why people use energy is not always well understood, even by the companies that sell it. In the 1980s, one California electrical utility recognized that its customers were using their refrigerators more than they had expected. Because they could not understand what was going on from a functionalist model of the refrigerator as a machine for food preservation, they hired anthropologists to investigate. They found that refrigerators were used for much more than storing food. People also used them to hide money in fake plastic cabbages, to allow pet snakes to hibernate, and to preserve photographic film, nylon stockings, and drugs. And at times people opened refrigerators with no definite purpose, mentally foraging, trying to decide if they were hungry or whether anything inside was appealing. Often they closed the door without removing anything. Because of such unanticipated uses, the refrigerators were using more power than had been expected. The anthropologists recognized that every mechanical device suggests new possibilities. Indeed, the refrigerator began "to take on altogether new identities—as a vault, closet, display case, morgue." On the basis of this study, the California utility was better able to predict demand and, to some extent, to change it. Likewise, every technology has multiple, unexpected uses, and citizens manage new products in visual, tactile, and verbal ways. Humanist methods not only can help others to understand this process; they also can suggest clear alternatives that are environmentally sound.

Yet reducing demand, either by improving appliances or by modifying the behavior of consumers, is only part of the solution. At the same time, energy production must shift away from oil, coal, and gas toward renewable energy. Harnessing the energy of the wind, the sun, and the tides can both reduce air and water pollution and reduce emissions of greenhouse gases. The problem is not primarily technical. In Germany, houses already have been built that are so well insulated and have such good heat exchangers that they are considered "zero-energy"—that is, to have zero net energy consumption. Likewise, windmills and solar power panels already are efficient and inexpensive enough to replace fossil fuels as sources of power, and the difficulties of storing electricity have been overcome more than lobbyists for coal and oil are willing to admit. In 2015 Denmark produced 42 percent of its electricity with windmills alone, and its combined renewable energy sources produced more than half. The following year Portugal produced almost half its electricity from wind and solar power. Yet the European Union as a whole generated only 10 percent of its electricity from solar (3 percent) and wind (7.5 percent), because some countries lagged far behind.[12]

As these statistics suggest, the problem of energy is primarily political and cultural, and it will not be solved by scientists alone. Theda Skocpol, a professor of government and sociology at Harvard University, has been quoted as saying, "If it were money [from lobbyists] only, it would be so much easier to deal with." But special interests are not the entire story: "Everybody on the left thinks it's only money and it's only Exxon. If it were, you could strike a bargain. It's definitely ideology, along with the usual kind of industrial lobbying."[13]

The ideology of energy remains deeply embedded in Western culture. That ideology has been challenged since the 1970s, but despite the widespread recognition of the dangers of global warming it remains intertwined with patterns of consumption and widely accepted economic paradigms. Solving the problem of energy is central to stopping global warming, to protecting many habitats threatened by energy extraction, and to creating a sustainable way of life. Energy is ultimately inseparable from the issue of consumption.

Consuming the Planet?

Humanists, bioscientists, and political scientists and economists typically mean quite different things when they use the term "consumption." An important shift in the use of that word occurred in the nineteenth century. Around 1800, most goods were still handmade, and the producer had a reputation, not a brand name. Sales were primarily local. In the countries that industrialized, by the early twentieth century goods were increasingly mass produced and marketed under brand names. The personal relationship between producer and customer was being rapidly replaced by a more abstract relationship between a corporation and a consumer. Buyers of goods increasingly relied on advertising for information. All too often, this new mass-production economy emphasized owning large quantities of goods, with no thought at all for the environmental impact this might have. For most of human history, for example, the average person owned only one or two pairs of shoes. But mass production of shoes brought their price down, and affluent people began to own different shoes for gardening, for walking, for going to work, for playing tennis, or for the beach. It became common in wealthy countries to own ten or more pairs of shoes, and the rest of the wardrobe expanded as well. There were environmental costs associated with the production, the distribution, and the eventual disposal of all this clothing. Where once people had few goods, and these often lasted for a generation or more, more and more goods were replaced simply because the owners were bored with them or because styles had changed.[14]

Many humanists took up the study of consumption in the 1980s, often reacting against the idea (once widely accepted by Marxists and capitalists alike) that consumers were easily herded sheep whom corporations controlled through advertising.[15] Scholars searched for examples of consumers making choices and of communities that used consumption to shape or reinforce their identities. For example, Lizabeth Cohen's early work on ethnic groups in Chicago in the 1930s showed that different communities purchased different goods in local stores, played different songs on radio stations, and otherwise expressed themselves through what they purchased and rejected.[16] Later, in a book titled *A Consumer's Republic,* Cohen traced the transformation of economic policies and described a political culture that increasingly came to view society not as a republic of sovereign citizens, but as characterized by atomized niche markets and sovereign consumers. In the consumer's republic, then, a logical political strategy is to market campaign messages to enough target demographic segments to win an election.[17] Similarly, a logical political response to worsening economic inequalities at home or wars abroad is for leaders to encourage citizens to shop more, thereby stimulating economic growth. As these examples suggest, consumers' choices take place in a complex political and cultural framework that calls for descriptive, historical, and interpretive research.

Advertising is hardly straightforward communication; it employs irony, humor, and bold parody to capture consumers' attention. In 2015, in collaboration with the artist Attila Hartwig, Volkswagen released a satirical advertisement for a purportedly gasoline-inspired perfume in Berlin during that city's Fashion Week. The ad showed a photo of a small black perfume bottle photographed against a backdrop of multicolored swirls of oil spilled on asphalt. Volkswagen also produced a slick mock TV ad identifying the perfume as "Mémoire de pétrole." Although the perfume is said to be "unisex," the ad features stereotypical images of objectifying masculine lust that fuse desire for women and auto bodies. A male voice speaks in an intimate tone of the smell of gasoline as "der Duft der Freiheit" (the scent of freedom). A monochromatic montage cuts from the outline of a naked woman to a clicking odometer and, just in case we missed the joke, ends with an image of a dripping twentieth-century petrol pump. Viewers are invited to laugh at the psychosexual attachments of an era obsessed with big gas-guzzlers and to indulge in nostalgia. The campaign was launched as satiric publicity for an all-electric Volkswagen model. In an additional, unintended irony, the ad campaign aired only a few months before it was revealed that Volkswagen had systematically installed software on its diesel vehicles that understated how much they polluted the atmosphere.

In both Europe and the United States, the word "anthracite"—literally the name of coal's darkest, hardest incarnation—has been used since the early years of this century to market a range of products and materials, including men's suits (Stellson), sports cars (Audi), women's purses (Balenciaga), and interior house paint. The paint may have been intended to match the dark granite countertops and stainless steel appliances that were then much in vogue for the upper middle-class. But the dark aesthetics of carbon are nothing if not democratic. In 2007, the mid-brow clothing retailer Old Navy, perhaps capitalizing on affluent Euro-Americans' attachment to the era of cheap and abundant fossil fuels, launched a line of T shirts featuring "vintage" logos of oil companies.

Successful corporations have become more and more adept at understanding consumers and making strategic accommodations in product design or in advertising to keep their loyalty. Before the end of the twentieth century, broadcasting was giving way to "narrowcasting" based on a sophisticated differentiation into market niches. Likewise, magazines that tried to speak to everyone were dying out, replaced by many specialist publications. The development of the Internet accelerated this process, allowing corporations to track individuals' preferences and send them messages based on their past consumption patterns. Within this larger context, it was not long before environmental issues came to the forefront of discussion. If economic orthodoxy teaches that "consumer choices drive the market," critics who study politics, culture, and society through qualitative, interpretive methods have observed how, in some sectors of the market corporations shape and drive consumers' choices.

In one unavoidable area of consumption—food—there is considerable evidence that agricultural subsidies, food industry marketing, and policies that make junk food widely available in public spaces shape food choices and create long-term patterns that link food, health, and lifestyle. This "food environment" and not an abstract "market" effectively constrains consumers' choices—a point repeatedly made by the nutritionist and public health scholar Marion Nestle in a series of books, beginning with *Food Politics*.[18] The cultural critique of food consumption, particularly as manifested in the "Western diet" and popularized by Eric Schlosser, Michael Pollan, and other authors, has stimulated an international public discussion about how the food industry influences consumers' often unhealthy choices. Some environmentalists have urged consumers to eat less meat or stop eating it altogether, because of the high environmental demands of raising animals, slaughtering them, processing the meat, and getting it to market. Some food choices have a much larger environmental impact than others, and consumers have become more interested in buying locally produced foods. There has been a slow-burning resistance to supermarkets and industrial agriculture from grassroots organizations that range from farmer's markets to cooperatives to consumer groups.[19] Many restaurants now make a point of serving locally produced food. Indeed, the market niche for "locally sourced," "artisanal," and "small-batch" products is so well defined that its well-heeled aficionados have become the object of satire.

One contribution of the environmental humanities is to make clear normative arguments: to describe the world not merely as it is, but as it could be. Another is to reveal ways of thinking that may seem to be merely descriptive but are in effect prescriptive. Because consumption is a matter of identity and cultural values, choices are not made in a neutral marketplace, but are shaped by many factors beyond advertising. If we are to live sustainably, these values must change, and one of the most influential ideas reshaping the behavior of consumers is that of the "ecological footprint." A country's "ecological footprint" is calculated by dividing its resource demands by its total population. To live within the limits of the earth's resources in 2012, each person should use no more than what can be produced on 1.7 hectares. By this measure, India is an exemplary country, with a footprint of only 1.2 hectares per person, while China's level of consumption is twice what it should be, at 3.4 hectares per person. Really drastic overconsumption, however, is found in Germany and the Netherlands (each with an "ecological footprints" of 5.3), in Slovenia and Switzerland (each with 5.8), and in the United States and Canada (each with 8.2).[20]

The environmental humanities can help make sense of historical patterns of consumer choice and help to develop more sustainable behavior. They can have a practical value for a government agency that wants citizens to adopt a new recycling system, or for a corporation that wants to develop a mobile app that will help consumers to make efficient use of its product. For students and citizens, the environmental humanities offer a salutary reframing of such choices as moments to exert thoughtful collective agency. Rhetorical, historical, and ethnographic analysis that is independent of market pressures is a valuable resource for studying and improving the behavior of consumers.

Intercultural studies, moreover, can help correct the false normalization of consumption patterns of countries in the global North. Many well-intended international development projects seek to bring "healthier" and "greener" technologies, such as improved stoves, into rural homes. Yet their efforts are often met with resistance. The Canadian researcher Njoki Wane found that Embu women in Kenya resisted giving up inherited utensils and hearths for less polluting solar stoves. The boxy solar stoves made preparing traditional foods more difficult and offered none of the body-warming heat of a woodfire.[21] And what if it turns out that rural villagers in Uganda prefer slightly less efficient stoves made in Kampala to more costly "eco-stoves" that promise a rebate from distant international organizations?

If consumers lack confidence that sustainable products have real benefits, they may reject them. A noxious cynicism can take hold that views all discussion of sustainable consumption as "greenwashing," dishonest marketing, or snobbism. Likewise, a proliferation of eco-labels produces a degree of confusion. Lack of critical, independent research on consumption could delay a transition to more sustainable consumption. More generally, consumers need to learn how to escape the Jevons paradox, in which increased efficiency only leads to increased consumption, with no overall savings. This occurs when clothes washers become more efficient, but consumers buy more clothing and wash it more frequently; or when televisions demand less electricity, yet consumers buy larger sets.

The environmental humanities can be useful in the educational campaigns that will be increasingly important if any country's carbon footprint is to diminish. Scientists reported in May 2013 that carbon dioxide had passed beyond the threshold of 400 parts per million. According to a report in the *New York Times*, "the best available evidence suggests the amount of CO_2 in the air has not been this high for at least three million years, before humans evolved, and scientists believe the rise portends large changes in the climate and the level of the sea."[22] Dealing with the "new normal" of climate change means dealing with consumption as part of larger collective problems. For example, in 2016 the World Bank warned that the world was unprepared for the disasters that will accompany climate change and will put 1.3 billion people at risk by 2050.[23]

The Belgian philosopher Isabelle Stengers has described consumption as a glaring contradiction at the heart of the world economic system that is driven by endlessly rising consumption on a finite planet with destabilized climatic conditions. This inconsistency undermines the conventional story that defines progress as consumption-driven growth. According to Stengers, exhortations to consumers to "have faith in growth" but at the same time "measure their ecological footprint" reveal precisely the "irresponsible and egotistical character of this mode of consumption."[24] Stengers further claims that "our leaders are responsible for managing what we could call a 'cold panic'"—a panic in which openly contradictory messages are accepted: "consume, because growth depends on it," but "think of your ecological footprint"; "recognize that our way of life is going to have to change," but "don't forget that we're engaged in a competition upon which our prosperity depends."[25] Faced with the impasse of a "cold panic" of a precarious economy on an ever-hotter planet, Stengers suggests that citizens demand more of leaders—in some cases, that they demand new leaders.

The dominant discourses of economics, management, marketing, and behavioral psychology bear much responsibility for the current impasse. In contrast, philosophical commitment to thrift and frugality was a central economic pillar to living a good life for thinkers from Diogenes to Thoreau to E. F. Schumacher. Minimizing consumption is not an elitist notion but is embraced by many groups devoted to voluntary simplicity and reduced consumption. They reframe consumption so that it is not a utilitarian, functional notion. The political theorists Tom Princen and Michael Maniates argue that the academy must likewise offer alternatives to the supply and demand curves and calculations of mainstream economics, which has treated consumption as a black box in its model of development.[26] Environmental humanists pose critical questions about consumption, rather than ratifying governing assumptions such as that society needs more growth or the misapprehension ("distancing" in Princen's terminology) of resource consumption as "production."

The environmental humanities offer many useful approaches to consumption. Citizens need to understand the life cycle of products, and see how long-term patterns of consumers' behaviors are related to mobility choices (When and why do urbanites choose to go into debt to purchase cars instead of bicycles?), food choices (To what extent is eating more meat a nutritional choice or an assertion of status?), and the energy intensity of shelter alternatives (Why are single-family detached houses so common in North America but rare in the Netherlands?). Environmental humanists study such contrasts and identify historical alternatives and paths not taken. The Indian historian and sociologist Ramachandra Guha highlights a related question that the economist John Kenneth Galbraith posed the same year as he published *The Affluent Society*: "How much should a person consume?" Scholars in the global North who focused on "environmental problems" long neglected this question, instead focusing on neo-Malthusian concerns about population growth. Yet the matter of consumption was at the center of the social justice and proto-environmentalist agendas of Lewis Mumford and Mohandas Gandhi. Another approach is suggested by anthropologists who have explored consumption along contact zones between cultures. Frictions occur around desires for goods and resistance to their world-changing impact on more circular, ecosystem-oriented livelihoods. Why might an indigenous community in Oaxaca whose material culture is based on fishing a lagoon using small sailboats reject a wind-power project pushed as "green development" by the provincial government and its transnational partners?

Human and cultural geographers question the choices made by consumers at every link in the commodity chain, from extraction to disposal. How do sought-after goods become undesirable waste? Cell phones are made in part from rare earths extracted in Africa and plastics from Middle Eastern oil. They are assembled in southeast Asia, purchased and used all around the world, and then sometimes exported as e-waste to Africa and China again. Critical geographers and others also analyze the spatial dimension of consumption, from the cathedrals of consumerism (megamalls, superstores, sports stadiums, festival tents) to austere refuges of behaviors on the edges of the spectrum of consumption (fasting, usufruct, upcycling).

Scholars of religions have studied the history, belief systems, rituals, and material life of voluntary simplicity, both among monastic orders that emphasize ascetic practices (Tibetan Buddhist saints, Franciscan monks) and among reformist non-conforming religious communities (Shakers, Hutterites, and many others, from Latin America to Japan). These groups demonstrate in practice that reducing consumption is not merely possible but also potentially rewarding as it shifts attention away from defining the self through possessions. Efforts to embrace a simpler life could draw on the principled plain living of the monastic tradition, or could learn from the Mennonites who continue to farm using horses and reject most high-energy technologies. A successful return to low-tech forms of agriculture is celebrated in the work of Wendell Berry.[27] Such a return might also draw inspiration from the writings of Henry David Thoreau and Aldo Leopold. Asceticism in the Western tradition stretches back to both ancient Greece and early Christianity. Plato describes Socrates as a man with few possessions whose greatest pleasure lay in philosophy. In the ideal society of his Republic, the philosophers display exemplary moderation. Jesus praised charity and alms giving and declared that it would be harder for a rich man to reach heaven than for a camel to pass through the eye of a needle. Some early Christian communities shared their possessions, and the monastic tradition emphasized living with a minimum of worldly goods. Monks in several Buddhist traditions maintain similar practices. Nor are luxuries common in Thomas More's *Utopia* (1516). Its citizens live in Spartan simplicity, wearing sturdy clothing whose fashion never changes. Because they have few possessions, More's Utopians labor only six hours a day and yet are never in want. Thomas Merton, a Christian monk whose books sold widely, praised the ascetic, meditative tradition. He declared: "'It is precisely because we are convinced that our life, as such, is better if we have a better car, a better TV set, better toothpaste, etc., that we condemn and destroy our own reality and the reality of our natural resources. Technology was made for man, not man for technology."[28] Merton was one of the first to read Rachel Carson's *Silent Spring*, and he corresponded with her about the misuse of technology and the "despair in the midst of 'plenty'" as hallmarks of a society of consumption.[29]

Were present-day societies to imitate More's Utopia, a redefinition of politics and economics would be required. In psychological terms, there would have to be a radical shift away from ownership and display of goods as sources of identity. In terms of the Kaya Identity, this would mean a severe reduction in GDP. The long-term historical trend is moving in a different direction, however. For centuries people have been moving off the land into cities. Even if as many as 100 million people were to return to a simple rural life and embrace low consumption, most of humanity would still be in cities.

Reconceiving Cities as Ecologies

During the Industrial Revolution many cities treated their immediate environment as raw material to be manipulated and transformed. Boston, for example, moved tons of stone and dirt to fill shallow bays and marshy areas, vastly increasing its size. For centuries, the Dutch expanded their country's land area by building dikes, controlling rivers, and building a complex canal system. In such cases, society was understood less as part of an ecological system than as humankind's imposition of culture on nature. But the Dutch no longer act or think that way. They have recognized that their cities are inseparable from the larger landscape and that there are quite literal limits to growth. In fact, Dutch researchers (among them Ruth Oldenziel, who coordinates an international research network on cycling and urban mobility) are at the forefront of projects in the environmental humanities whose purpose is to create a new urban ecology. The world's cities keep growing, and the United Nations estimates that they produce 75 percent of the world's carbon emissions. Cities' demands for food, water, energy, and waste removal are at the center of the environmental crisis. Yet, at the same time, cities offer opportunities. In the twentieth century, they became field sites for ecological research[30] and hotbeds of insurgent place-making,[31] guerilla gardens, night markets, and environmental activism. As the UN put it, "the high density of cities can bring efficiency gains and technological innovation while reducing resource and energy consumption."[32]

Before cities industrialized, they were smaller. Commonly a city contained many gardens, and it was not unusual to raise chickens in back yards.[33] Much of the food consumed in cities was produced nearby. But industrialization transformed food supply as canning, refrigeration, and steamships made it possible to transport grains, fruits, vegetables, and meats halfway around the world. This was energy intensive and environmentally costly, and it promoted monocultures and specialized production. Enormous districts grew only wheat or corn, while other areas specialized in milk, butter, and cheese. To some extent, specialization allowed for the most economically productive use of land. But there were also cases in which local populations went hungry while nearby land was devoted to producing for export. Does it really make sense to export beef from Namibia to Denmark, itself a major meat exporter? Probably not.

Consuming locally produced foods is not only about economics. It is also an important part of the effort to recover a sense of, and a feeling of responsibility for, place (discussed in chapter 2). If residents understand a city as an ecology, then they will want to patronize restaurants that rely on nearby farmers. Even if they do not have the option of growing their own food, they may choose to recover the practice of home canning that was common from about 1870 until after World War II. They can also support initiatives to reduce their community's pollution—for example electric buses, better home insulation, efficient lighting systems, and bike lanes for commuters. Such initiatives are piecemeal measures based on the realization that human beings are not outside nature. Human beings are biological organisms, and their cultural institutions are tightly linked to natural cycles and the landscape.

The understanding that cities should be seen as ecologies is by no means new. It was embraced by the Chicago School of sociology in the 1920s, and even earlier by the University of Chicago's department of geography. Though the ecological model then employed had undertones of social Darwinism if not racism, an analogy was insistently made that emphasized similarities between plant, animal, and human communities. In recent years a movement has developed to transform cities after again conceiving them as ecologies. Some city planners have returned to the idea of urban agriculture, albeit in new formations. In line with that way of thinking, many buildings now have "green roofs" that save energy and reduce rain runoff, and architects design new structures that minimize energy use. The new generation of urban planners wants to reduce a city's CO_2 emissions to zero. This is the goal of the cities of Sønderborg in Denmark and Växjö in Sweden. Between 1993 and 2011, Växjö reduced its CO_2 emissions per inhabitant by 41 percent. In 2013 Växjö hosted a meeting on "Energy-Cities," and it has partnered with a number of Southeast Asian cities.[34] Such communities have moved beyond piecemeal reforms to plan for large, systemic change.

Cities are more ecologically defensible than suburbs. David Owen argues in *Green Metropolis* that cities already have developed ways for people to live closer together, reduce their driving, and minimize their pollution. The most inefficient living is in the suburbs. In comparison with the residents of Long Island, the average New Yorker lives in a smaller space, drives much less (or not at all), and has a smaller carbon footprint. Because row houses and many other apartment buildings share walls, they demand less heating than a house. In short, cities usually make more ecological sense than suburbs. City residents, on average, use only half as much gasoline, electricity, and water as people who live in the country.[35] Moreover, they can use mass transit and bicycles more and cars less. They can redesign streets and sidewalks to make walking safer and more inviting. They can recycle more, in less wasteful refuse systems. City governments can change the building code. A glass-walled building gets plenty of light and thereby saves electricity, but on sunny days it overheats, requiring expensive ventilation and cooling systems, and on a cold night even the best argon-filled triple-paned window loses much more heat than a well-insulated wall. The roofs of some buildings, apartment terraces, and allotment gardens collectively can supply a significant fraction of a city's food, and at the same time reconnect people to the cycle of production.[36] However, as Owen takes pains to demonstrate, the total energy use in producing apples or lamb in New Zealand and shipping them to London or New York may be less than the energy needed to produce the same goods much closer to the consumer. It is counter-intuitive, but consuming only locally produced foods may not always be the most environmentally appropriate choice. The carbon footprint of a New Zealand lamb turns out to be only one fourth that of one from Scotland.[37]

There is more than one route to creating a sustainable city, not least because every city must be adapted to its local environment. Let us examine some specific cases. In Abu Dhabi the government has invested billions of dollars in Masdar City, a project endorsed by the World Wildlife Fund. It is to have a dense population and to be powered by solar energy. Thanks to careful design, it is expected to use only one fifth as much energy as a conventional city. No automobiles will be allowed; there will be electric mass transport beneath the streets. It is too soon to judge whether Masdar City will achieve these goals. It may turn out to be an artificial oasis of solar virtue in the midst of generally wasteful use of fossil fuels in the Middle East, or it may become a model for others to follow in building new cities in desert climates. Some of Masdar City's features may be transferable to other areas, but in most cases sustainability will be achieved not by constructing new cities but by retrofitting existing ones.

A good deal of the writing on redesigning existing cities into sustainable ecologies advocates their transformation into "smart cities" that will use digital systems to increase efficiency and improve the sharing of resources. However, in a book titled *Sharing Cities* Duncan McLaren and Julian Agyeman suggest that such efforts can all too easily be co-opted by a gentrification agenda in which poor, heterogeneous communities are pushed out to make way for new, higher-income residents. This process, which Sarah Dooling calls "ecological gentrification," has occurred in Seattle, New York, London, and many Western cities, though it has also been successfully resisted—notably, as McClaren and Agyeman write, in Medellin, Colombia, where new cultural institutions and parks were placed in poor neighborhoods as a stimulus to their improvement.[38] McClaren and Agyeman also cite Bangalore, India's third-largest city. In the years 2001–2011 its population grew by 47 percent to almost 10 million, creating crises in garbage collection, traffic, electricity supply, and water supply. With its many IT professionals, Bangalore may solve these problems and become a "smart city." The danger, as Richard Sennett has noted, is that "smart cities" can become heavily monitored and controlled from a central command center. They can become "over-zoned, defying the fact that real development in cities is often haphazard, or in between the cracks of what's allowed."[39]

Sennett and other close observers understand that cities are dynamic and that change energizes competing groups. The architect Steven Moore has studied efforts to achieve urban sustainability and has found that their advocates divide into factions. In Austin, Texas, "green romantics" think that the way "to change the world is by changing the consciousness of individuals, one at a time, by altering their aesthetic experience of nature."[40] In contrast, "economic rationalists" seek technocratic solutions, such as clean energy or more efficient homes produced by private firms or "smart cities." "Rugged individualists," who dominate Austin's politics, expect free markets to solve ecological problems. Each of the various groups in Austin that seek to influence governmental decisions tells a different story, and success hinges on making a narrative convincing.

Moore also examined the German city of Frankfurt, where the long-dominant political ideology was based on liberal capitalism, maximizing individual freedom in the marketplace, though it also recognized the complexity of natural systems. From this perspective, the solution to environmental problems was "soft" technological determinism, in which engineering subordinated nature, with environmental protection from state agencies. However, the Green Party in Frankfurt developed a powerful alternative argument. It won control of the city in 1989, and during the following six years its politicians worked from quite different assumptions. They assumed there were global limits to growth, rejected technological determinism and argued instead that technologies are socially constructed. They rejected mechanistic conceptions of nature as raw material to be subordinated and controlled, and instead had organicist views of human beings as immersed within nature. However, as Moore emphasizes, this was not a simple opposition between liberals and radicals. Each group had its own factions. There were "green romantics" and "green rationalists." Some progressive capitalists emphasized ecological modernization through planning; others trusted the market to find solutions. In all, Moore identified seven different kinds of discourse, which together defined a complex dialogue. It resulted not merely in compromise but in the creation of new building codes that reimagined the new skyscrapers then being built in the city center. "The American model of a skyscraper," Moore writes, "was not simply appropriated ... and then redecorated with new architectural signs to render it German, or green-washed with a few solar collectors to render it sustainable. Rather, this building type was reimagined through the social construction of new technical codes." As the result of their dialogue, "Frankfurters were able to imagine an alternative future."[41]

Unlike an experiment built from scratch, such as Masdar City, existing cities, like Frankfurt, must evolve into new kinds of communities. And through community-based design, they will seize upon ideas that planners did not foresee. For example, solar panels are now being installed on hydroelectric plants' reservoirs and on lakes at water-treatment plants. This has several benefits. The water cools the panels, while the panels reduce evaporation and inhibit algae growth, both common problems on sunny lakes. A Japanese power company deployed on a reservoir 50,000 floating panels that can supply 5,000 homes with power, as well as preventing evaporation of water used for hydropower.[42] As this case illustrates, environmental solutions can discover new uses for a site that make it more productive. Such solutions arise not from science, technology, engineering, and mathematics operating in a vacuum, but through consultation and inspiration from studies of culture, social interaction, and the fine arts and design fields.

Urban mobility is another area in which humanists can contribute to making cities more sustainable. Historians, anthropologists, and urban theorists are collaborating to try to understand how urban citizens choose among forms of available transport as well as the long-term cultural politics that condition present-day choices. An international team of researchers across the environmental humanities is exploring past "cycling cities" for the purposes of engaging policy makers and clarifying a vision of sustainable urbanism. They are examining historical traffic data and using ethnographic methods to bring visibility to street-level experiences in relation to urban policy and traffic engineering.[43] How did cars come to dominate so many cities? Why were popular, accessible, and low-consumption methods of mobility such as pedestrianism and cycling coded as "old-fashioned" or "dangerous," despite often contrary evidence? An international dialogue in the environmental humanities takes mobility as a nexus of energy, consumption, and urbanism and offers new ideas to cities from Shanghai to Helsinki.

The movement toward sustainability is more successful when there is dialogue between groups than when pre-conceived solutions are imposed. The environmental humanities can improve that dialogue, in order to discover a narrative of change and to negotiate the technical codes that make sustainability possible. As the Kaya Identity makes clear, there is an imperative need for such dialogue in order to halt population growth, improve energy efficiency, and reduce the use of fossil fuels. Otherwise, the ice in Greenland and Antarctica will continue to melt, the seas will continue to rise, and many of the world's cities will experience periodic floods. The urgency of this and other environmental crises has led some scientists to advocate the global engineering projects discussed in the following chapter.

Figure 4.1
Gene Daniels, "Oil Waste on Barren Hillside, Coalinga, California, 1973." Documerica Series, Record Group 412, 542512, National Archives.

4 Promises and Dangers of Science

The pace of scientific development has quickened during the last two centuries. Until roughly 1850, small groups of scientists and tinkerers made occasional discoveries. By the late nineteenth century, efforts were more organized. Entrepreneurial inventors began to establish permanent teams of researchers, perhaps most famously at Thomas Edison's research laboratories. In 1900 General Electric established the first dedicated corporate research laboratory in the United States, and other large businesses soon established laboratories as well. Their goals were both to improve existing products and to discover entirely new ones, creating an endless stream of patents that would ensure corporate dominance in a particular industry. DuPont invented nylon, for example, and IBM developed patents for mainframe computers. Corporations usually sponsored applied research, while universities and research hospitals performed more of the "blue sky" or pure research without regard for its possible practical applications. But such divisions are not absolute. Some corporate researchers won Nobel Prizes for their work, and some university researchers set up profitable corporations. In both the private and the public sector, the number of researchers has exploded since 1921, when just 3,000 scientists and engineers were employed by US corporations. Eighty years later, their numbers had increased to 1.2 million. The OECD countries had 3.3 million researchers in 2002, and their work accelerated the pace of change. Many discoveries and inventions had profound implications for humanity's place in the natural world. This chapter will examine three topics closely related to these developments: biotechnology, the rising rate of species extinction, and proposals for large-scale modification of the earth and its atmosphere through geo-engineering.

Biotechnology: The Science of Small Things

After the discovery of DNA and the development of laboratory equipment that facilitates its isolation, identification, and replication, it became possible to manipulate genetic codes in order to create new drugs or to modify the genome of plants, viruses, animals, and human beings. The potential benefits and dangers are enormous. New techniques of more precise gene editing developed since 2010 have revolutionized the scope of research; biologists, bioethicists, and science councils are hastily trying to draw new guidelines for work that raises fears of a revival of eugenics or of inadvertent public health crises.[1] In 2015 there were 1,947 international journals of biochemistry, genetics, and molecular biology,[2] in which scholars could present the miraculous capabilities of their research well before it became legal to develop and sell the results in the marketplace.

Biotechnology in agriculture promises to feed a population of 10 billion with genetically modified drought-resistant and disease-resistant grains redesigned to provide more complete nutrition. Science grant writers may compete for funding with which to create new forms of life, resurrect extinct species, end chronic diseases, and enhance human intelligence and longevity. One genetic engineering company has found a way to change a leopard's or a dalmatian's spots. It is working on a redesign of the angus cow so that it will be not black but white and thus able to thrive in warmer weather. To demonstrate the technique, the same company has already produced mice with either square "spots" or straight lines in their fur.[3] For investors, biotech startups represent the next big opportunity on a par with the dot-com boom at the turn of the millennium.

Until recently the promises of biotechnology seemed overblown, sounding more like science fiction than laboratory science. Then researchers completed the first human genome sequence in 2001. In the next ten years, computing power increased, data sharing among labs accelerated, and the computational and biological sciences converged in subfields such as bioinformatics. Advances in nanotechnology, electronics, and imaging supported explorations of highly complex phenomena. Interdisciplinary efforts are now underway to unlock the mystery of consciousness. The BRAIN (Brain Research through Advancing Innovative Neurotechnologies) initiative of the National Institutes of Health aims to map human brain activity down to the millions of neurons that constitute circuits of thought, emotion, and behavior.[4] In 2016, the editors of *Nature Biotechnology* predicted that "all biology will become computational biology." Genomics, the sequencing and analysis of complete sets of cellular DNA, requires enormous computing power. Analyzing genes' functions requires "megabases" rather than mere databases; it also requires transnational computational centers linking multiple sequence-analysis engines and genomic data warehouses. Projects such as the Oak Ridge National Laboratory's GRAIL (Gene Recognition and Assembly Internet Link) have made some areas of biology "big data" sciences on a par with astronomy and physics.

Molecular science and genetics are big businesses. There is even a managerial term that governments and foundations use when discussing applied biological research: "the bioeconomy." New startup companies have few qualms about inventing new life forms. It is a part of their business model. As one close observer of the industry commented, "The founding principle is to promulgate synthetic biology by making it easier to adopt."[5] One way to make it easier has been to reduce the cost of making DNA. Where in 2002 it cost about $10 to write a DNA pair of two "letters," by 2016 it cost only 3 cents. However, there are thousands of "letters" in even a short strand. Production of DNA is still costly, and a full human genome still has not been produced. In contrast, "a human cell that divides makes a new genome in 24 hours, basically for free."[6] Yet the speed of change resembles the exponential growth in computing power after 1960, and the (re)production of DNA is now being industrialized.

The revolution in biotechnology has raised utopian expectations. For example, George Church, a Harvard Medical School geneticist, has said that scientists soon will be able to recover extinct species. Yet he also sensed that "our technological capacity outstrips what it all means" and asked "Who will be doing this and what are the regulations?"[7] The environmental humanities are sorely needed to frame such questions of ethics, politics, and history in the public realm alongside the framings of natural scientists and those of leaders of civic and religious organizations. Indigenous rights groups, farm activists, religious conservatives, and traditionalists have not been silent on the manipulation of genes, the appropriation of living tissues as intellectual property, and the hubris of humans assuming God-like powers. Secular humanists active in the fields of bioethics and medical ethics have had less to say about the broader environmental impacts of biotech.

Suppose the passenger pigeon were to be brought back, for example. Millions of these birds once flocked across North America, at times consuming the crops of Euro-American settlers, who likened them to a plague of locusts—and hunted them into extinction as an easy source of protein. Passenger pigeons functioned within a radically different ecosystem: as primary consumers of tree fruits, they spread beechnuts and the acorns of once-dominant white oaks. Once the great clouds of pigeons were gone, white oaks gave way to red oaks across the temperate forests of North America. If passenger pigeons were to be revived, should they be let loose throughout their former range? In what numbers? Researchers suspect that their reproductive success and survival depended in part on the size of the flock. Now that it is becoming practically as well as theoretically possible to revive species whose DNA is available from stuffed specimens, what ethical and political guidelines should guide de-extinction and rewilding? The environmental humanities, and especially new studies in philosophy, history, anthropology, and literary criticism, are engaged with bioscientists in updating ethics for the new data-driven wave of genetic engineering. Not content to rehash the critique of science associated with postmodern critical theory of the 1980s and the 1990s, a new generation of humanities scholars are investigating the justifications, the emotional grounding, and the aesthetic creativity at play in biotech.

The Canadian anthropologist Natasha Myers, for example, depicts protein modelers in her book *Rendering Life Molecular* as artisans passionately and viscerally involved in their work to discern and represent ("render") the structure of the proteins they study. To explain why a protein model is wrong, a modeler contorts her arms behind her back; students who prepare proteins for imaging speak of getting their samples "happy" and "relaxed." Computational software renders beautiful visualizations of folded, colorful ribbons of protein structure. The very beauty of these visualizations and the power of modeling technology may deceive students and even peer reviewers, who must be trained not to confuse beguiling models with real proteins. Myers acknowledges that bioengineering tends to commodify life: "Protein structures are becoming objects of multidisciplinary interest and investment. With the promise of novel insights into basic biological processes, biomedical research, drug development, biofuel engineering, and environmental remediation, biologists, chemists, physicists, engineers, mathematicians and computer scientists are accessing the coordinates of the vast array of structures housed in the Protein Data Bank. Value can be extracted from this data in the form of patentable designs and innovations."[8] Part of the story of microbiological engineering, she observes, is about bringing new products to market. In the hands of bioengineers, "proteins are rendered as the 'machinery of life,' and this machinery can be reengineered, repurposed, and 'enterprised up.'" "Life itself," it seems, has been captured and put to work in the form of a "streamlined assemblage of molecular machines that hum productively on the factory floor of our cells."[9] Pharmaceutical companies, for example, seek to build "molecular machines" to deliver targeted treatments for cancer and other illnesses. The microbiological assembly line might one day stretch from inside our bodies to beyond the boundaries of our planet. NASA is currently investigating the potential of astrobiology, particularly the search for evidence of the simplest forms of life such as bacteria on other planets in our solar system, and exploration of Mars is on the docket for the twenty-first-century.[10] Soil samples gathered during space exploration will be scrutinized for biological indicators of ore deposits in the regions where they were collected. Enhanced microbes could be used to "biomine" copper or uranium. Sulfur-loving bacteria can replace conventional methods (heat, pressure, chemical leaching agents) and perhaps extract gold or copper from sulfide formations more cheaply and with less toxins downstream. Biomining is in part inspired by efforts to develop microbes to clean soil and water in post-mining landscapes and oil spills.

Bioengineering at the molecular level is also of interest to national military agencies and to industries that produce weapons. If redesigned proteins can unleash (or perhaps leash) human potential, then understanding neural circuits, protein structure and function grants a total, embodied power. The French philosopher of science and historian Michel Foucault called this form of power "biopolitics" and characterized it as the power to grant life or to withdraw the conditions of life. Foucault associated the rise of the modern state in the nineteenth century with a strategic shift from systems of power organized around ritual execution (putting citizens to death) to biopolitics organized around institutions such as hospitals and the sanitary practices of public health policies. By the 1990s, humanities and social science researchers in Europe and in North America were adapting Foucault's concept of biopolitics for use in a critique of the overlap of technoscience and capitalism, militarism, and neo-colonialism. Since then, biopolitics has reached down to the molecular level.

Yet a blanket critique of biotechnology misses the complexity, the creativity, and the public spirit of much work in this field. Scholars can develop more nuanced narratives about technobiology. As the "Big Science" of small and vulnerable things, biology and its sibling discipline ecology appeal to humanists oriented to environmental problem solving. Anthropologists, philosophers, cultural critics, and historians are in productive conversations with biologists working on such diverse problems as developing more bountiful strains of maize, gene therapies for Alzheimer's, or oil-eating bacteria to clean up polluted coastal wetlands. Myers has called for "other analytic frames and other ways of telling stories about the sciences and lives in science" rather than the often-repeated reductive story of "capturing life" for profit.[11] Using an ethnographic method, Myers dramatizes the complexity and ambivalence of laboratory life. Her observations of graduate students and professional crystallographers reveals a range of personalities, interpersonal dynamics, and a "life affirming" biopolitics that "keep[s] open what it is possible to see, say, feel, and know about both scientific practice and the stuff of life."[12] Again and again, rather than assuming that modelers view proteins as cold, mechanistic matter, she finds them enlivened by visualization, metaphor, and "intimate relationships" with "their molecules."[13] The aesthetic and emotional richness of scientists' work questions the dominant narrative about science as mastery or control of nature.

The environmental humanities need to allow room for such a critical relation to the life sciences: "If observers of science just follow the scripts that scientists think they are supposed to follow, we would aid in entrenching and normalizing the hubris of stories of capture. We would be complicit in limiting the kinds of inquiry, and the modes of attention and relation that are possible in the life sciences. In this sense, stories told about science by those of us observing from a distance, which frame science as the capture of 'life itself' for capital gain, risk reproducing the very conditions that constrain what scientists can say, feel, imagine, and know."[14] Instead, the environmental humanities might adopt the paradigm of critical friendship with the life sciences.

By returning to deceptively simple questions that bridge epistemology, ontology, and life sciences, the humanities can connect the whiz-bang promise of biotechnology with the persistent challenges to realizing its implicit moral claims to human and environmental betterment. These challenges lie more often in culture and society than in the laboratory, but humanists and bioscientists can build on a strong cross-disciplinary inquiry in the areas of bioethics, environmental justice, and philosophy of science. What is the stuff of life? Why are so many people hungry? Similar questions are raised by advances in genetics, such as cloning, genetically modified crops, and synthetic biology.

Calls to publicize and share biological datasets notwithstanding, biotechnology as a business sector depends on patenting ideas—and its staunchest critics argue that it is intent on patenting life itself.[15] So under what conditions may biological technologies be considered public goods? How should the resulting knowledge be shared or distributed? Which biotechnologies are unethical or pose security threats? Such questions have become more urgent because the falling cost of DNA modification enables amateurs to begin tinkering with plant and animal biology. In the spring of 2013 a group of hobbyists successfully raised a quarter of a million dollars for a project to add genes for bioluminescence to trees, which they expect one day might replace streetlights. Despite the unforeseeable and potentially dangerous effects of haphazard introductions of new transgenic species, popular scientific magazines celebrated this as "a sustainable alternative to electric illumination."[16] At the very least, ethicists and historians should examine the potential effects of such biotechnologies before they are introduced, partly through comparisons with the impacts of past species introductions.

Radical environmentalists, artists, and writers of speculative fiction have been among the first explore the ramifications of such biotechnology. In *What We Leave Behind*, the "green" anarchist thinker Derek Jensen acknowledges the contradiction in accepting advanced medical treatment for cancer while writing anti-civilization tracts. In her MaddAddam trilogy, Margaret Atwood imagines synthetic bioforms run amok in a dystopian near-future world ruled by a sinister alliance of military contractors and pharmatech corporations who develop super-intelligent pigs and weaponized genetically modified dogs until a vengeful gene coder designs a super-pathogen to wipe out the human species. Then, under a domed conservatory reminiscent of Biosphere 2, the amoral gene hacker designs non-violent, leaf-eating humanoids to replace humans.

Taking a critical stance toward biotechnology does not mean that the environmental humanities turn their back on insights drawn from advances in biology. In fact, large ethical and political questions have emerged from within the biological sciences. The theories of the biologist Lynn Margulis challenged what was taken for granted about the emergence of life on this planet. She argued for a view of life as emergent from symbiosis rather than competition, and as rooted in cooperative, multi-species coexistence rather than monopolistic, species supremacism. This worldview begins with a radical egalitarianism: "All beings alive today are equally evolved," Margulis wrote in the introduction to one of her books.[17] This view of life rejects human exceptionalism and has influenced environmental ethics, theories of posthumanism, queer ecology, and the multispecies turn in history and anthropology. (See chapter 7 below.)

Biodiversity and Extinction

In 2016 a scientific report prepared by experts from Canada, the United States, and Mexico found that since 1970 the number of birds in North America had declined by a billion.[18] Of the 1,154 known species, 432 (37 percent) are considered to be at risk of extinction. Their habitats are shrinking, they face invasive predators, and the climate is changing. The species with long migration patterns have lost 70 percent of their population. There were isolated success stories, usually because governments and private conservation groups made a special effort. The problem is not lack of knowledge, but too little political and social will. Unfortunately, this grim report is not unusual. The species diversity of most areas on the earth has been declining for 200 years. Many species, genera, and orders (e.g., amphibians) survive in reduced numbers and genetically homogenized populations. Salmon disappeared from many rivers as a result of pollution or dam building, for example, while surviving in other areas. In the case of salmon, there has been some success in restocking rivers that had lost them, but other species disappeared entirely, in all or parts of their range. Island and endemic species, such as the extinct Costa Rican golden toad, have been particularly plagued by extinctions. These are losses that affect entire ecological systems in ways that are difficult to grasp using statistics alone.

Even as biological losses mount, scientists are still assembling the list of the species that do exist. Roughly 18,000 new species are discovered each year. By this measurement, the golden age of taxonomy—of discovering, cataloging, and naming earth's diverse forms of life—is now in full swing.[19]

Such astounding variety itself poses a challenge that the natural sciences alone cannot answer: how can one raise sufficient interest in new varieties of slime mold or wasps to ease the pressures on these creatures' habitats? When human niche construction worldwide threatens to wipe out the thousands of niches inhabited by other creatures, threats to biological diversity call for cultural, social, and philosophical approaches.

Historian Mark Barrow draws a similar conclusion about past conservation efforts in *Nature's Ghosts: Confronting Extinction from the Age of Jefferson to the Age of Ecology.* Conservation-minded naturalists in the early decades of the twentieth century often found themselves studying creatures threatened by many factors, and yet they were unable to prevent their numbers from dwindling further:

In the late 1930s, the National Association of Audubon Societies established a graduate fellowship program to research the life history and ecology of endangered species. The first two Audubon-sponsored projects—James Tanner's ivory-billed woodpecker study and Carl Koford's California condor study—uncovered a wealth of information about the habitat needs of and the threats facing their respective subjects. Both naturalists also offered a series of conservation recommendations consistent with those findings. In the end, though, neither succeeded in reversing the ongoing decline of their study subjects. *Science might be necessary to rescue endangered species, they discovered, but it was hardly sufficient.* What was ultimately needed was a broader political, social, and cultural shift to support additional research and to implement the often costly management recommendations that scientists made.[20]

Tanner and Koford, like many other scientists, were both specialists and public voices during the dawning age of ecology. Their efforts led to a greater political and cultural willingness to confront extinctions in the 1960s and the 1970s. These efforts coalesced around the term "biodiversity" by the late 1980s, and in 1992 it was enshrined in the UN Convention on Biological Diversity (CBD). Extinctions and ecological damage persisted, however, despite official recognition of biodiversity as a priority by more than 190 countries. Recognition of the problem did not mean that oil and mining companies, for example, immediately became more careful with wastes.

A large part of the problem was that the language of "biodiversity" had been captured by the reigning ideology of "development." Biodiversity's history is shot through with disabling contradictions; one of which is that the United States, a prime mover in global economic policy, has refused to sign the main international policy instrument, the CBD. Another contradiction inheres in the policy itself, which recognizes biodiversity as a source of value *for* social and economic development. Here is a symptomatic sample of the official language on biodiversity from the website of the CBD: "The Earth's biological resources are vital to humanity's economic and social development. As a result, there is a growing recognition that biological diversity is a global asset of tremendous value to present and future generations. At the same time, the threat to species and ecosystems has never been so great as it is today. Species extinction caused by human activities continues at an alarming rate."[21] This doublespeak treats threatened species as fungible assets in the global market for ecosystem services. Playing value-laden terms off one another provides ample wiggle room for governments, corporations, and individuals. Species and ecosystems become both biological resources to mine and diversity to be conserved. What bars powerful countries from plundering their weaker neighbors? Instead, governments need to recognize and reward cultural practices that enhance diversity in ecological systems as a whole, rather than take a species-by-species approach that lends itself to seeing each life form as an asset or resource.

Species diversity and debates over the categorization of species, from Darwin's work to Ernst Mayr's philosophy of biology, are important areas for humanities research. It may be salutary to recall Mayr's view of the internal complexity of disciplines and of the broad connections between biology, history, and narrative. "There is more difference between physics and evolutionary biology," Mayr wrote, "than between evolutionary biology (one of the sciences) and history (one of the humanities)."[22] Outside of the rather narrow preserves of science, other cultures have long perceived diversity as a part of ecological knowledge. Anthropologists and political ecologists have explored non-Western awareness of the diversity in ethnobotanical and ethnographic studies. Similarly, after 2014 the Indonesian government sought ways to ground species preservation in local and religious values.[23] One could imagine government scientists working alongside anthropologists in Muslim communities in West Sumatra to preserve rainforest through faith-based initiatives, for example. Widespread human perception of and appreciation for species diversity precedes the taxonomic systems of official Western science, and in many places religious beliefs shape conservation decisions. Conservation may be a non-starter in cultures if theology is not part of a larger conversation with researchers and government.

In this comparative light, the term "biodiversity," introduced by E. O. Wilson and other conservation-minded biologists in the late 1980s, can be seen as a "scientization" of diversity in the West, another "big idea" in the tradition of large-scale funding in science. As a new object of research, biodiversity became an eligible object of policy and governance. At the 1992 United Nations Conference on Environment and Development in Rio de Janeiro, biodiversity officially became a discursive tool in geopolitical relations through the Convention on Biological Diversity. The literary critic Cheryl Lousley has observed how, in the writings of Wilson and other prominent biologists and in UN policy they inspired, the term "biodiversity" became almost synonymous with rainforests.[24] From the beginning, discussions of biodiversity became entangled in power relations between the global North and the global South. In legal terms, this has been articulated as the obligation of policy makers to distribute costs equitably between more and less developed countries. The relatively small Central American country of Costa Rica exemplifies the problematic political-economic relations that meet in the term "biodiversity." Costa Rica has set itself the goal to stop using fossil fuels by 2022. However, its ecotourism sector depends in large part on marketing the country's natural beauty and flying in international birders to see its many rare endemic species.

Nineteenth-century observers were aware that local extinctions of "useful species" of game and fish had occurred during their lifetime. In *A Week on the Concord and Merrimack Rivers* (1849), Thoreau noted that he had witnessed the disappearance of migratory fish on Massachusetts rivers:

Salmon, Shad, and Alewives were formerly abundant here, and taken in weirs by the Indians, who taught this method to the whites, by whom they were used as food and as manure, until the dam, and afterward the canal at Billerica, and the factories at Lowell, put an end to their migrations hitherward ... . It is said, to account for the destruction of the fishery, that those who at that time represented the interests of the fishermen and the fishes, remembering between what dates they were accustomed to take the grown shad, stipulated, that the dams should be left open for that season only, and the fry, which go down a month later, were consequently stopped and destroyed by myriads.[25]

A Week on the Concord and Merrimack Rivers weaves natural histories and personal histories into an autobiographical travelogue retracing the stages of a canoe trip Henry David Thoreau made with his brother John in August 1839. The local extinctions visible in Thoreau's lifetime shaped a brooding sense of loss. He offered a hyperbolic lament to the shad: "Away with the superficial and selfish phil-*anthropy* of men,—who knows what admirable virtue of fishes may be below low-water-mark, bearing up against a hard destiny, not admired by that fellow-creature who alone can appreciate it! Who hears the fishes when they cry? It will not be forgotten by some memory that we were contemporaries."[26] Such a misanthropic outburst was unusual in mid-nineteenth-century literature. Perhaps what made *A Week* a publishing flop when it appeared makes it nonetheless more interesting now, for we live among an untold number of species that face an uncertain future, "at the dull edge of extinction."[27]

Even as researchers continue to discover new forms of life in the deep seas and in boreal forests, the rate of extinction is very high, and climate change is a major threat.[28] Since Precambrian times, there have been five spikes in the fossil record where a large percentage of known species have disappeared, the best known of which is the last extinction of the dinosaurs and mesosaurs about 66 million years ago. Globally, habitat destruction and new species introductions have led biologists and geoscientists to describe what is now underway as a sixth global extinction event, as was briefly described in chapter 1. Environmental humanists attend to the losses of meaning and culture that accompany this loss of ecological diversity. Others have calculated the potential loss of "ecosystem services" or speculated on possible rainforest cures for cancer that could be lost to science. Humanists ask "What does the loss of megafauna as well as less visible forms of life mean for how we define ourselves?" This is an ethical, onto-existential, political, and (for some) theological question. Are we to be the God species, not only naming but choosing which creatures live and die? What does it mean for salmon-fishing cultures or whale-hunting cultures to witness the extinction of the species on which they have long depended and the diminishing of their material lives, their rituals, and sacred myths? Without the whale and the salmon, what becomes of the Maori whale rider or the Salmon People of the Columbia River Basin? Local and regional species diversity are inextricably linked to social, political, and normative issues, whether such diversity is explicitly tied to a particular culture ("biocultural diversity"), is documented in impact assessment studies, or is the primary object of conservation. Historians, philosophers, ecocritics, and anthropologists as well as conservation biologists have conceptualized biological diversity and its protections. They have, for example, reframed its ethical challenges and the political and social contexts in which such diversity is negotiated.

The innovative practice of "multispecies ethnography" offers a way to understand human activities alongside and in parallel with the agency of other still-evolving forms of life.[29] Such scholarship de-centers humans as sovereign producers of knowledge with a monopoly on rights. "If we appreciate the foolishness of human exceptionalism, then we know that becoming is always becoming *with*," as Donna Haraway writes in *When Species Meet*, building on the work of Isabelle Stengers and Gregory Bateson.[30] Ethicists have taken extinctions and losses of other forms of life as the primary challenge of this century. In *Flight Ways*, Thomas van Dooren describes five avian species at the brink of extinction and the many people caring for them, including those caregivers who seem locked into what he calls regimes of "violent care."[31] Such care may be justified, but its darker side should not be forgotten. For preserving threatened species often comes at a high cost to individual animals. Not only are competing introduced species exterminated, but captive breeding programs themselves pose challenges for how we treat the Hawaiian crows, Galapagos tortoises, or whooping cranes we aim to preserve. Breeding whooping cranes, for example, has involved lifelong incarceration of surrogate brood mothers from their biological cousins, the sandhill cranes.

The everyday mixture of interspecies care and violence is more apparent on farms and in other rural working landscapes. Mauro Agnoletti, a historian of forests and landscapes, has argued that culture can also be a boon to biodiversity. With a team of researchers, Agnoletti produced a national guide to rural landscapes in Italy. In addition to producing a set of rich categories and descriptions of cultural landscapes (a category of protection often associated in Europe with heritage areas), Agnoletti found that biodiversity was greater in historically significant agricultural landscapes than in subsequently reforested or abandoned areas.[32] This insight resulted from combining a historical perspective with vegetation ecology and broad geographical knowledge. A textured picture emerged of how areas changed over time in concert with human activity. However, anthropologists and geographers expressed skepticism about Agnoletti's claims to biocultural diversity almost as soon as they emerged (circa 2001), because the hotspots of species loss are also long-term inhabited landscapes in Brazil, Zambia, Indonesia, and Madagascar. More research is needed.

In summation, a "background rate" of extinctions is a normal part of evolutionary processes, but the rate of species loss due to human activity is far higher, and can be considered a mass extinction that has been unfolding for the last 20,000 years. The enormous increase in the human population since 1800, coupled with industrialization, has accelerated the rate of this extinction. Indeed, it is possible that humanity itself will so overtax the earth's resources and its capacity to absorb pollution that it leads to self-destruction.

Geoengineering: A Whole Earth Experiment?

The term "geoengineering" refers to intentional large-scale modifications of earth systems to solve problems such as global warming, acidification of the oceans, or desertification. Such projects have entered the realm of possibility in recent times, owing to the greatly enhanced technological capabilities of modern societies. Grandiose plans that once would have been the stuff of science fiction emerged as serious proposals in the twentieth century. In the 1920s the German architect Herman Sörgel proposed to build an enormous hydroelectric dam across the Straits of Gibraltar that would lower the level of the Mediterranean Sea and open new land areas to colonization. In 1949 a group of Stalin's engineers advocated an equally grandiose project, the Davydov Plan, which proposed to dam two Siberian rivers so that instead of running north into the Arctic Ocean they would form a lake four times as large as the Aral Sea whose water could be sent through a 2,500-mile canal to fertilize deserts to the south. Such Soviet river diversion projects continued to be discussed in the 1960s and the 1970s.[33]

While historians continue to investigate the "prehistory" of such large dams, canals, and tunnels as precedents for futuristic schemes to control the weather (a sensational topic favored by the Soviet and American press in the 1950s) or for the use of nuclear explosives to create a sea-level canal linking the Pacific and the Atlantic, the term "geoengineering" itself is profoundly futurist in its orientation. In the twenty-first century, as evidence of global climate change becomes more apparent and threatening, geoengineering has become a lightning rod for criticism. Today it would be possible, for example, to use enormous pumping stations to divert seawater into the Sahara Desert, which would increase the humidity there and at the same time counteract the rising levels of the seas due to the melting of the ice sheets on Greenland and Western Antarctica. But how would such a project affect humans and other species living in the desert? Would the evaporation of water eventually create a vast salty soup that could not support aquatic life? Or would the introduction of so much water fundamentally change the climate of North Africa by lowering temperatures and increasing rainfall? Even if the most optimistic scenario were valid, would it be ethically defensible? For example, would such engineering be politically desirable if it could help island societies threatened by rising seas? Similar questions must also be raised about the solar radiation management (SRM) projects that have already received millions of dollars in support. These proposals raise the issue of whether humans should respond to global warming by reducing their demands on the environment or by seeking to modify the weather.

Here we will not attempt to consider all the geoengineering projects that have been dreamed up. Rather, we will focus on geoengineering techniques that are both typical and probable. These proposals have the backing of scientists—that is, credible authorities—even if the ideas themselves have not yet been subjected to peer review. Scientists have applied for funding for projects to place thousands of small mirrors in orbit around the earth to deflect sunlight and thereby reduce global warming. Such a scheme, if it worked, might lower temperatures too much, posing the expensive problem of removing the mirrors. Other scientists are attempting to reduce the amount of methane produced in the stomachs of cows. (Methane, when released into the atmosphere, increases global warming.) How re-engineering bovine digestion might affect the surrounding environment is not clear. Other scientists propose to increase the ocean's ability to absorb carbon dioxide by adding crushed limestone or processed pelletized lime across thousands of kilometers of the Antarctic Sea, by fertilizing plankton blooms, or by seeding the water with iron filings.

The geochemist Paul Crutzen concluded that SRM technologies were worth further exploration and debate and discussed chemical means of "artificially enhancing earth's albedo [heat-reflecting capacity] and thereby cooling climate by adding sunlight reflecting aerosol in the stratosphere."[34] Crutzen himself had the expertise necessary to design experiments on the atmospheric chemistry of stratospheric sulfur dioxide—an "aerosol" that could, in the manner of a major volcanic eruption, increase the earth's albedo. Yet, as the social theorists Alexander Stoner and Andony Melathopoulos point out, the presence of sulfur dioxide in the lower atmosphere precipitated horrific public health crises such as the 1952 Great Smog of London. Atmospheric sulfur once practically defined air pollution, and it was rightly understood to be a global problem requiring coordinated regulation. "Today," Stoner and Melathopoulos write, "sulfur presents itself to us not as a constituent part of the planet to be regulated by society ... but as an inevitability, as something that we will collectively be forced to shoot into the stratosphere because of our inability to take hold of the runaway character of society itself."[35] How can governments and citizens negotiate the expert discourse that dignifies geoengineering proposals such as SRM?

This is not the first time that scientists have proposed to "fix the sky." The historian of technology James Rodger Fleming has described past ambitions to control climate. Some past efforts were the province of charlatans as well as military researchers interested in developing new weapons—including the US military's Operation Popeye during the Vietnam War, which sought to increase rainfall in areas held by the Viet Cong.[36] Fleming's work is an object lesson in why philosophers, historians of science, and environmental historians should be included in considerations of both the possible benefits of geoengineering proposals and their unintended consequences and indirect costs. There is a heavy-handed interventionist bias in the designs of geoengineering projects. The idea seems always to be that an expensive technological intervention is the cure. A strong drive to fund geoengineering in the style of the Cold War is evident in the calls from some national science agencies for "more research" projects that promise to mitigate climate change. The involvement of agencies (including DARPA) that historically have been interested in the potential of weather and climate manipulation as a military weapon gives both cause for alarm and fodder for geoengineering conspiracy theories that have proliferated online. The Royal Society in the United Kingdom and National Academy of Sciences in the United States have published detailed summaries of the challenges and possibilities associated with re-engineering the atmosphere to combat climate change. Yet, as the philosopher Clive Hamilton has pointed out, these nuanced policy documents might be used to justify risky manipulation of planetary systems with heavily unequal outcomes—especially in the event of a "climate emergency" such as a rapid rise in sea level.[37] Hamilton has explored the ethical and political dimensions of geoengineering, which, he argues, "is not just a technological problem, nor even an ethical one as usually understood; it goes to the heart of what it means for one species to hold the future of a planet in its hands."[38]

A few prominent environmentalists have been among the voices suggesting that geoengineering is inevitable. Jedediah Purdy, writing in the digital magazine *Aeon* in October 2015 shortly before the publication of his book *After Nature*, asserted that "the conversation about climate change has shifted from whether we can keep greenhouse-gas concentrations below key thresholds to how we are going to adapt when they cross those thresholds. Geo-engineering, deliberately intervening in planetary systems, used to be the unspeakable proposal in climate policy. Now it is in the mix and almost sure to grow more prominent."[39]

What are the stakes associated with how we *talk* about geoengineering, whether we describe it as "inevitable," "a necessary risk," or as "dangerous," "a sign of hubris?" The jargon that has been invented to describe geoengineering—Climate System Intervention, Novel Options, Earth System Engineering and Management, Planetary Manipulation, Climate Remediation, Climate Stabilization, Negative Emissions Technologies, Carbon or GHG Removal, Carbon Geoengineering, Carbon Capture and Sequestration, Radiation Management, Sunlight Reflection Methods, Solar Geoengineering, Targeted Climate Modification, Albedo Modification, Reflectivity Modification—has the potential to obscure its benefits and its dangers. Critical discourse analysis is catching up with the inflated but rather indefinite language of "planetary management." The contrast between the relatively cautious language of the Royal Society's 2009 report on geoengineering and the more bullish, interventionist style of the National Academy of Sciences' reports on "climate intervention" is an early indicator of how prominent governing bodies are lending credibility to geoengineering and adapting their rhetoric to soothe concerned citizens. Independent experts without a direct financial stake in geoengineering research projects are underrepresented in funding agencies and on national research councils.

The humanists who are best able to provide an informed historical and ethical perspective on geoengineering have responded with genuine alarm. James Rodger Fleming, for example, has questioned whether "management" is even an appropriate designation for schemes such solar radiation "management." As early as 2009, he declared "Global climate engineering is untested and untestable, and dangerous beyond belief."[40] The ethicist Stephen Gardiner has found the arguments for deploying technological fixes to mitigate a climate emergency worse than unconvincing. They lead, he writes, to "horrifying moral territory"—for example, victimized island nations are forced to beg the very countries responsible for rising seas to deploy geoengineering to "save" them. Gardiner reportedly resigned from one government review panel because he felt that the group assigned to assess the social value of geoengineering was giving short shrift to its ethical complexity.[41]

In view of the increasing efforts to justify research, let alone to deploy geoengineering, we should heed the voices of ethicists, historians, and political theorists, and the diverse concerns, human and nonhuman, represented by civil society and indigenous groups. The "science" (that is, formalized, probabilistic forms of knowledge) of geoengineering relies on too narrow a band of geophysical, geochemical, and economic expertise. The public, before adopting or rejecting specific interventions in the global climate, should hear from researchers in the fields of linguistics, ethics, literature, cultural anthropology, political and social theory, and history of science. Not everyone agrees that a technological fix can solve global problems, and in many cases the "cure" may be worse than the problem.

Wordplay and clever marketing have thrived in the absence of public attention to humanistic and social science study of geoengineering. Two reports published by the National Research Council in 2015 (one titled *Climate Intervention: Carbon Dioxide Removal and Reliable Sequestration*, the other titled *Climate Intervention: Reflecting Sunlight to Cool Earth*) used the term "climate intervention," which perhaps sounds more benevolent than geoengineering to the public. Given the complexities involved, the international group of geoengineering experts is shockingly small. Moreover, these researchers often have a patent or other personal stake in the methods proposed, such as patents for carbon extraction. Some of these researchers have received funding from the Bill and Melinda Gates Foundation. Bill Gates, the founder of Microsoft, has enthusiastically endorsed inventing new technologies to extract carbon from the atmosphere or to block solar radiation—technologies that when patented and deployed could produce planetary path dependency on "proprietary technology." The profit potential may have been calculated more carefully than the potential for environmental damage or injustice.

At the very least, geoengineering raises serious questions about governance, including whether any existing institutions are adequate to govern its implementation.[42] Might one country pursue SRM or a similar project without the consent of other countries' governments? Could a rogue entrepreneur fertilize a swath of ocean to increase snowfall in a region in which his vast real estate holdings rely on winter tourism? Do Samoans, Senegalese, Maori, Anishinabe, and Laplanders want to live in a climate that has been debugged of overheating? Partial answers to these questions have already been given: Yes, Yes, and No. Environmental activists have already prevented two attempts by corporate entities to dump iron fillings in the ocean, once in 2007 and again in 2009. These planned dumpings defied a "moratorium" that had been declared by the United Nations after delegates to the 2009 UN Indigenous Peoples' Global Summit on Climate Change denounced geoengineering as a "false solution"—a judgment that was echoed by subsequent gatherings of indigenous leaders, including the People's World Conference on Climate Change and the Rights of Mother Earth (held in 2010 in Bolivia).

Artists are also engaging the ambiguous political ramifications and sci-fi appeal of geoengineering, and their engagement underlines how such projects require imagination and rely on crafting new worldviews. Kim Stanley Robinson's *Science in the Capitol* trilogy imagines a near-future world in which geoengineering schemes are pursued as national policy in North America. Media and visual artists have also responded to geo-engineering, including Karolina Sobecka, who in 2016 filmed an experimental performance of a low altitude cloud-making machine.[43] Bio-responsive interactive virtual games and a fictional scenario transparently titled *The Collapse of Western Civilization* imagine multiple partial failures to engineer our way out of a climate-altered world, not only by spraying stratospheric sulfur but also by genetically engineering a black lichen with super CO_2 absorbent potential.[44] The specter of a geoengineered world, perhaps more than any other technological fixation, has stimulated dark visions of collapse and catastrophe—including entire continents blackened by invasive lichens.

Toward the Posthuman?

Another scientific "solution" might be to create synthetic human beings with DNA engineered for a world transformed by climate change. That idea is not being funded, so far as we are aware, but 150 scientists did meet in a closed door meeting at the Harvard Medical School in May 2016 to discuss the possible synthesis of human DNA.[45] In other words, they think it possible not just to copy or to edit DNA but to create it from chemicals. The human beings that would result would have no parents, and presumably would be designed so as maximize health and intelligence. Such schemes begin to resemble the ideas explored in Aldous Huxley's dark satire *Brave New World,* in which people were designed and conditioned to be members of four distinct classes.[46] Once one discovers that we are already living in a world in which such technologies are being pursued, dark visions of the future begin to proliferate.

Scientific "improvement" of human beings is not new. Eyeglasses, hearing aids, artificial limbs, heart pacemakers, and many kinds of surgical implants have become routine, and at least in a small sense have begun to transform ordinary human beings into cyborgs. Decades ago, in an essay titled "A Cyborg Manifesto," Donna Haraway argued that the breakdown of barriers between the organic and the technological was creating new categories of experience in a post-gender world.[47] Katharine Hayles has argued that, as a result of a dissolution of the firm distinctions between nature and culture, we are entering a posthuman era.[48] In an interview, Hayles summarized an emergent view of human beings that is sharply differentiated from the Enlightenment view:

Whereas the human has traditionally been associated with consciousness, rationality, free will, autonomous agency, and the right of the subject to possess himself, the posthuman sees human behavior as the result of a number of autonomous agents running their programs more or less independently of one another. Complex behavior in this view is an emergent property that arises when these programs, each fairly simple in itself, begin reacting with one another. Consciousness, long regarded as the seat of identity, in this model is relegated to an "epiphenomenon." Agency still exists, but it is distributed and largely unconscious, or at least a-conscious.[49]

The posthuman, in short, sounds much like a cluster of semi-autonomous computer applications—a view that Hayles does not accept. Human beings are embodied and live in places, not in cyberspace. But after about 1950, as information was reconceived as a disembodied flow that could be endlessly replicated and stored, and eventually uploaded to "the cloud," human beings were accordingly reconceived to resemble computers. And with R2D2 in the *Star Wars* films, the wetware cyborg in *Avatar,* and other representations, popular culture began to present robots as friendly and non-threatening. Attempts to elide the differences between human beings and machines, whatever their ideological premises or cultural location, promote a posthuman view of what a person is. If such views are accepted, then manipulating DNA seems to be nothing more than upgrading our hardware, and the transfer of consciousness to machines would appear to be a likely eventuality. Scholars of the environmental humanities may be divided on this question; however, the majority hold the view that human beings are not machine-like, that they are embodied, and that even if they find ways to enter cyberspace, as in some science fiction, attempts to go and live there permanently are likely to end badly.

Scientific proposals for reengineering human DNA, the body, the earth, the atmosphere, or outer space are themselves a form of technological socialization that accustoms people to radical intervention in the structures of the earth and the forms of life. The environmental humanities interrogate such radical interventions, recognizing that some of them might be necessary or useful, while maintaining a critical distance from technological fixes and looking for less drastic alternatives. To many, geoengineering, species extinction, and artificial DNA suggest not progress but dark visions of the Anthropocene.

Figure 5.1
"Mass Extinction Memorial Observatory," artist's rendering, Adjaye Associates. The observatory is currently under construction on the Isle of Portland, England where it will exhibit carved images of every extinct species recorded since the passing of the dodo.

5 The Anthropocene's Dark Visions

A World of Wounds

Aldo Leopold wrote that "one of the penalties of an ecological education is that one lives alone in a world of wounds." The alternative is to remain oblivious to the signs of damaging human activities, which in Leopold's day might still pass unnoticed. But now, each week brings fresh news of disasters: fierce storms linked to anthropogenic climate change, droughts that affect whole continents, and entire ecosystems such as coral reefs and primary rainforests threatened by extinction. Planetary damage and disparate harms to human health are more salient than in the middle of the twentieth century, when Leopold developed his land ethic. Leopold saw a stark choice: "An ecologist must either harden his shell and make believe that the consequences of science are none of his business, or he must be the doctor who sees the marks of death in a community that believes itself well and does not want to be told otherwise."[1]

This chapter concerns the discovery, definition, and interpretation of the Anthropocene, including dark visions of planetary collapse. The Cassandra cries of ecological disaster echo through dystopian novels, films, nonfiction accounts of actual disasters and the struggles to survive in scarred, damaged lands.

Planetary ecological crisis also bleeds into the gothic horror of popular TV shows. A US series titled *The Walking Dead* follows a well-armed, violent band of survivors after a "zombie apocalypse" wipes out most of humanity. While not explicitly an environmental-themed show, its post-apocalyptic world resembles the one feared in the late twentieth century by neo-Malthusians and sketched in more recent projections by the Intergovernmental Panel on Climate Change. The planet is haunted by extinctions, food shortages, violent conflicts and other effects of exceeding its carrying capacity.[2] In short, global climate change and other incremental, long-term forms of environmental damage are manifested directly in the visual arts, in literature, and in popular culture, feeding a new surge of pessimism, fatalism, and the macabre.

Anti-environmentalist critics, some funded by free-market think tanks, have loudly denounced the "doomsayers" of environmental degradation. Such critics deny links between human activities and climate change.[3] Yet carefully documented historical trends of species extinction, global warming, and population growth pose real dangers.[4] Visions of ecological crisis are not merely a "dark patch" to avoid in favor of more pragmatic or rational responses to the Anthropocene. Their imminence, scale, and probability remain uncertain; hence the prognostic value of speculative and even apocalyptic fictions. The concept of the Anthropocene gathers many dismal megatrends under a single banner; it also serves as a kind of philosophical wrecking ball. It dismantles and delegitimizes promises of economic growth based on clever engineering and on Adam Smith's idea of an "invisible hand" guiding the free market. Moreover, the tendency of ecological education to evoke grief, anger, resentment, and other disturbing emotions justifies renewed study of culture, literature, history, the arts, and philosophy.

The concept of the Anthropocene announces both the possible end of human life and the need for an alternative way of being in the world. Roy Scranton describes this as "learning to die in the Anthropocene":

If *Homo sapiens* survives the next millennium, it will be survival in a world unrecognizably different from the one we have known for the last 200,000 years.

In order for us to adapt to this strange new world, we're going to need more than scientific reports and military policy. We're going to need new ideas. We're going to need new myths and new stories, a new conceptual understanding of reality, and a new relationship to the deep polyglot traditions of human culture that carbon-based capitalism has vitiated through commodification and assimilation. Over and against capitalism, we will need a new way of thinking our collective existence. We need a new vision of who "we" are. We need a new humanism—a newly philosophical humanism, undergirded by renewed attention to the humanities.[5]

The humanities can guide these processes of mental ripening and ethical learning. Indeed, the environmental humanities have developed precisely in response to the threat of a possible ecological collapse, and their tasks include studying the alternatives, including dark, ironic, and satirical responses. These explorations help to define necessary action rather than succumb to analytical paralysis. *Contra* Leopold, an education in the environmental humanities does not leave one alone in a world of wounds, but among a new multitude. For even as the Anthropocene names a planetary human environment as an object of concern, the singular "human" at its center is fundamentally ambiguous. Dark visions of ecological crisis and social strife are a symptom of the age of humans as a planetary force. Awareness of the possibility of abrupt, widespread unpleasant changes—from a resurgence in lethal contagions to financial panics driven by speculation on limited critical resources—motivates a new ethics, a new politics, and new ways to imagine being human.[6]

This chapter first frames the science, history, and controversy surrounding the Anthropocene before introducing related work on environmental disasters, particularly on the slow violence of mining, nuclear disasters, ecosickness, and climate change and on how to reconstruct life after loss. These studies are an important part of environmental education, moving readers through denial, anxiety, fear, and anger toward knowledge and action.

Naming the Anthropocene: Science, History, and Controversy

Like evolution, the concept of the Anthropocene was developed independently by two scientists: the biologist Eugene Stoermer, who studied diatoms as indicators of past environmental change in the Great Lakes, and the geochemist Paul Crutzen, whose utterance of the term "Anthropocene" at a geological meeting at the turn of the millennium prompted Stoermer to join forces with him. Stoermer and Crutzen announced the term as a proposed new geological epoch in May 2000 and framed the Anthropocene as a grand challenge for humanity: "To develop a world-wide accepted strategy leading to sustainability of ecosystems against human induced stresses will be one of the great future tasks of humankind, requiring intensive research efforts and wise application of the knowledge thus acquired."[7] Before discussing the politics of this concept, we note that there are two distinct strands of evidence. The geological strand is rooted in the lithosphere. It peels back the strata of human influence and looks back to the history of mining and the geochemistry of carbon in the world's oceans. The geophysical account of the Anthropocene might lead us to see geoengineering as a natural extension of the Age of Man. In this new order, geochemists could develop techniques to set the global thermostat to a preferable temperature (but preferable to whom?) and "fix" the climate. In contrast, the biological concept of Anthropocene is haunted by the shadow of the sixth great extinction. The disappearance of thousands of forms of life from the future fossil record has been hastened or caused by the expansion of agriculture, overfishing, deforestation, and other human actions. This would seem to justify renaming the Holocene or "new whole" as the era of humans. That the whole of a creaturely world could unravel in a geological moment as a result of human actions (intended or not) places the self-conscious intelligence of *Homo sapiens* under the sign of hubris.[8]

The Stratigraphic Commission, which decides official names for geological history, asked a special sub-committee to consider the evidence for renaming part of the Quaternary Period (all of the Holocene Era, roughly the last 12,000 years) or perhaps a subset of the Holocene Era (since 4000 BCE or after 1750 CE), or simply not changing the name at all. Looking through the Anthropocene lens, the entire earth becomes an artifact. Evidence includes artificial elements, neoliths such as rocks made of compressed human garbage, and trace elements once rare but quite common in modern life.[9] The geologist Jan Zalasiewicz is fond of citing the tungsten carbide in a ballpoint pen as a quotidian trace of the Anthropocene.[10] Many substances that now are widespread did not exist before twentieth-century synthetic chemistry.

Other traces of global human activity include the worldwide layer of radioactive isotopes left from above-ground atomic weapon explosions in the years 1945–1960. Above-ground tests also generated centimeters-thick neoliths in localized deposits, such as Trinitite or "Alamogordo glass," the mildly radioactive glassy mineral left after the Trinity Test of 1945, which is now coveted by "rock hounds." Among still other traces of human activity are the largest manufactured landscapes on earth: landscapes where mining has leveled entire strata in areas of the southern Appalachians, Wyoming's Powder River coal region, Brazil, Southwest China, Ghana, and South Africa.

In addition to studying mines and neoliths, scientists analyze global climate change as a geological force. From the perspective of future geologists, a central question might be "When did human modification of earth systems reach a point where it would leave a noticeable trace in future strata?" Scholars have focused on three periods. The earliest is the Neolithic revolution, during which human societies felled and burned vast forests and began living in agricultural settlements. A second possibility is the industrial revolution that began around 1750 and altered atmospheric chemistry by pumping more carbon into the air. Yet some scientists focus on the "Great Acceleration" announced in 1945 by the atom bomb. Dating the beginning of the Anthropocene depends upon finding a reliable marker (a "golden spike") in the geological strata. This presents a temporal paradox: Although the half-lives of radioactive isotopes dwarf human time scales, they fall far short (by powers of ten) of the scale of most geological periods.[11]

What will scientists millions of years from now find in outcroppings from circa 2000 CE? Will they have instruments and techniques sufficiently precise to discern differences at the scale of centuries? They may find far less limestone than one would otherwise expect from a period of relatively high seas. Atmospheric CO_2 is absorbed by the world's oceans, where it binds with water molecules to form carbonic acid (H_2CO_3), thus lowering pH. These acidic waters impair sea creatures' ability to build shells (calcium carbonate). Because more acidic oceans reduce populations of shell-building creatures, far less calcium-rich sediment sinks to the ocean floor, which affects the amount and distribution of future sedimentary stone.

Future geologists are not likely to find evidence of glaciation during the Anthropocene. An article in *Nature* estimated that global warming will forestall the next ice age by at least 100,000 years.[12] Although that might seem a boon to our descendants, earth-shaping glaciers have some positive effects. Some of the world's most productive soils are a legacy of the last glaciation, benefiting agricultural yields by their high content of fine minerals. Glaciers alter the distribution of vegetation and fauna for tens of thousands of years, and they actually increase the supply of land as sea level falls. They cause local and even total extinctions of endemic species while benefiting other species. Thick ice also compresses the earth's crust so that, for thousands of years after the retreat of the glaciers, vast regions slowly "rebound," gaining elevation.

The historian Dipesh Chakrabarty has pointed out that global climate change signals a convergence of human and natural history.[13] Historians have also entered the discussion of periodizing the Anthropocene, and two of them—John McNeill and Naomi Oreskes—have joined the Working Group on the 'Anthropocene' of the International Union of Geological Sciences' Subcommission on Quaternary Stratigraphy. In 2016, that group seemed to be converging on dating the start of the Anthropocene in the middle of the twentieth century. But some humanists have strenuously objected to thinking of the "new human age" as a product of a dramatic rupture with the past. The historian Greg Cushman has analyzed the data sets for extraction of raw minerals, population growth, and other indicators from 1830 on. Closely attending to data on past mineral use and extraction (notably of phosphorus), Cushman hypothesizes that the Great Acceleration may in fact be a *second* acceleration. An even more rapid period of expansion in plantation agriculture and the number of livestock occurred between 1880 and 1910. In other words, one could reasonably date the Anthropocene not from 1945 but instead from the culmination of imperial expansion. The era's name more accurately might be "Plantationocene" or "Eurocene."

At stake is not only when the transformation began but who bears responsibility. The earlier date places greater responsibility for the Anthropocene on Europe and North America. Dating the change from the 1940s and using the prefix "anthro" spreads out responsibility to all human beings. Environmental historians have tended to adopt a date considerably earlier than the late nineteenth century. They have pointed to the impact of the Columbian exchange and decline of human populations across the Americas, which by 1620 certainly had extinguished tens of millions of household fires and begun a period of reforestation. This took a great deal of carbon *out* of the atmosphere. The Amazon, before Columbus, was not a wilderness; it was thickly inhabited.[14] Much of the depopulation of the Americas could not have been intended, as Europeans did not know about viruses or germs, but conquest was aided by what Alfred Crosby called invisible "ecological co-invaders."

If the debate in environmental history and among geoscientists turns on when the Anthropocene began, intellectual historians have pointed out that the concern with global human impact on the environment is not new. In 2002 Paul Crutzen cited the Italian geologist Antonio Stoppani's 1873 discussion of the human power to shape the globe as evidence of an "anthropozoic era."[15] But awareness that human beings were reshaping the natural world was widespread by 1770. Carl Linnaeus warned that dire consequences could follow the elimination of even one species of earthworm, French natural scientists were concerned about the decline of fishing stocks in the 1770s, and European foresters protested the deforestation they witnessed, fearing disruption of the weather, declines in animal populations, and the drying up of springs.[16] By 1904, the Austrian geologist Eduard Suess theorized a planetary geological system (the lithosphere) and corresponding layer of life, the "whole of the animal world" or "biosphere."[17] The Russian geochemist Vladimir Vernadsky lectured at the Sorbonne in the 1920s and introduced his version of the "biosphere" to a circle of French scholars that included the philosopher Edouard Le Roy and the paleontologist and theologian Pierre Teilhard de Chardin, who subsequently wrote of a "noösphere"—a world of human thought. Vernadsky came to view the earth as a system of interlocking spheres, including the lithosphere, the atmosphere, and the technosphere.[18] By the middle of the twentieth century, popular writers (among them Murray Bookchin and Rachel Carson) were noting that novel chemicals had penetrated living systems from the Arctic to the tropics and could slowly poison individuals and threaten entire species, as in the case of DDT's effects on endangered raptors.

The future orientation of the Anthropocene forces us to stretch our sense of the duration and scale of human impact into geological time, including ambiguous histories of unintended consequences and tradeoffs from megadams to horrific plagues to industrial agriculture to banning CFCs. Unlike Verdanksy's and Teilhard's formulations of the "noösphere," naming the Anthropocene does not denote an evolutionary leap forward. Rather it indexes dire material effects that human intelligence long failed to comprehend. Large-scale burning of fossil fuels occurred for about 150 years before anyone understood its atmospheric forcing effect. Nearly another hundred years passed before a consensus formed that greenhouse gases caused global warming.

Interpreting the Anthropocene

Interpretations of what was to be called the Anthropocene began well before the term was invented. They began with multiple observations of degradation across biophysical systems: overfished seas, clearcut forests, strip-mined and leveled mountains, encroaching deserts, dead zones at the mouths of major rivers, and mass extinctions and die-offs across whole orders of life (amphibians, marine mammals). The Great Acceleration after 1945 witnessed an astronomical growth in human population and consumption, and the story of global ecological crisis emerged through journalism and nonfiction works that invented a new language of environmental disaster. Indiscriminate, widespread application of synthetic pesticides amounted to biocide. Population overshoot and subsequent collapse were known to wildlife ecologists earlier in the twentieth century. Now the specter of global human population overshoot—a "population bomb"—was seen as a threat on a level with nuclear holocaust. Crop failures, droughts, and multi-year famines with suspected anthropogenic causes (deforestation and soil erosion) were identified as symptoms of a larger, looming question: Could human civilization be made safe for the planet?

Since the 1970s, some scholars have extrapolated from historical population trends to reach apocalyptic conclusions. In areas where the birth rate is 3.0 children per woman, such as rural India and much of Africa, the population was doubling every 24 years. The resource demands also doubled every 24 years. A population of 500 million in 1900 could increase to 8 billion at the century's end. That growth rate was unsustainable, and collapse seemed unavoidable. More recently, Rob Hengeveld, a Dutch specialist in resource depletion and population growth, concluded that "the collapse of the present human population, its numbers and quality of life, is likely" and that it will be the "unavoidable result of the behavior of an oversized, complex, nonlinear system in which interdependent chance processes dominate."[19] Hengeveld's faint hope is that governments will pre-empt this disaster by adopting policies that will reduce populations and hold consumption to a sustainable level. The choice is between uncontrolled collapse and downsizing to a smaller population. Downsizing will demand a radical transformation of human ideals, economic behavior, and political ideology. William H. Calvin reached similar conclusions in *Global Fever*, a book focused on global warming.[20] He concluded that unless governments reduce emissions of CO_2 and methane, rising sea levels will inundate the world's coastal cities, displacing hundreds of millions of people. Calvin also offered practical suggestions on how to cut emissions. But can voters and governments be induced to act? In a book titled *Down to the Wire: Confronting Climate Collapse*, David Orr argues that the world has entered a period of emergency in which decisive actions—including the relocation of millions of climate refugees, careful management of an agriculture system stressed by drought, unstable weather, and mediation of international conflicts over food and water—will be needed to avoid collapse. Orr claims that governments have confused growth with prosperity, that new economic models are needed to guide policy, and that different values will be needed if people are to live comfortably with less.[21] He calls for more honest leaders who will explain the environmental problems humanity faces and will inspire the public to embrace necessary change.

Revelation of humankind's planetary impact has been likened to a cosmological or theological event. The conceit that we are living in the Anthropocene or "era of humans" heralds the end of a known world and provokes feelings of existential peril. The German philosopher Peter Sloterdijk writes: "Everything suggests that we should construe the term 'Anthropocene' as an expression that makes sense only in the context of apocalyptic logic." Sloterdijk meditates on how the modernist acceleration of time proves inseparable from Martin Heidegger's concept of "running forward into death."[22] Haste took on a highly specific meaning once human beings began to see themselves not as inhabitants of a stable natural world with ample resources but as passengers on a fragile planet—a planet that R. Buckminster Fuller began to call "Spaceship Earth" as early as 1968. A dominant species is perversely destroying its life-support systems and the "planetary boundaries" for maintaining safe conditions for human civilization. Indeed, when an interdisciplinary research group proposed nine such boundaries in 2009, it concluded that human-driven changes have already exceeded three of them.[23]

Because scientific knowledge often diffuses slowly, people with environmental concerns often have relied on moral appeals to concepts that scientists no longer accept. Notably, they often appeal to ideas of equilibrium and balance, in the belief that those ideas are rooted in the science of ecology. They conceive of nature as an inherently conservative system that was in balance until humans disrupted it, and therefore they think that the goal of environmentalists must be to restore that balance. As Dana Phillips pointed out in *The Truth of Ecology*, much nature writing and many ecocritics were appealing to the "balance of nature" long after natural scientists developed a different view that emphasized how ecosystems change. Since the 1970s, ecologists have come to see biophysical systems as open rather than closed and as dynamic rather than stable. Biophysical systems may even change profoundly and rapidly rather than gradually.

Bill McKibben had the pristine, stable idea of nature in mind when he announced in 1989 had nature had "ended" as a result of anthropogenic climate change. Nevertheless, *The End of Nature* discerned many of the psychological and moral implications of Anthropocene discourse. Reviewing the science on global warming, acid rain, and a host of planetary-scale effects linked to human activity, McKibben concluded that "we live in a postnatural world."[24] That news took years to sink in, as he predicted.[25] Why were science writers such as McKibben and Andrew Revkin (who wrote of an "anthrocene era" in the 1990s) briefly celebrated and then widely ignored? An entire generation reached adulthood between the first awakening to the threat of global warming and the subsequent annunciation of the Anthropocene. And a decade passed after the stir in the natural sciences that happened around 2000 before historians and philosophers of science grasped its normative and epistemological resonance. There were outliers, of course. Elizabeth Kolbert mentioned Crutzen's Anthropocene argument in her three-part series "The Climate of Man" in *The New Yorker* in April and May of 2005, and Dipesh Chakrabarty's 2009 article in *Critical Inquiry* titled "The Climate of History: Four Theses" widened the ambit of concern.[26]

Responses to the Anthropocene range from a passionate, almost messianic embrace of the term to its outright rejection as a reinvention of Eurocentric false consciousness or as a revival of patriarchal anthropocentrism. Writings by the visionary philosopher and historian of science Donna Haraway, from *Simians, Cyborgs, and Women* to recent work on companion animals and multispecies co-becoming, have inspired many to ask question such as these: What is the nature of the human at the center of the Anthropocene? What, in addition to mourning the collapse of systems of intricate biological diversity, is to be done as citizens of a new epoch? Are humans, colonized as we are by friendly and fiendish microbes and enhanced by cybernetic prostheses, ready to act more generously to kindred beings? How many are ready to take up Haraway's call to "make kin" instead of more children?[27]

To what extent does the concept of the Anthropocene imply a normative framework (what humans should do or have done) masquerading as a universal politics? Chakrabarty described the Anthropocene as a "shared catastrophe that we have all fallen into,"[28] a vision of universal misfortune that the philosopher Slavoj Žižek and others have criticized.[29] As Rob Nixon has written, "We may all be in the Anthropocene but we're not all in it in the same way."[30] Does the Anthropocene concept jeopardize precious gains of expanded human rights and other Enlightenment ideals, including the Universal Declaration of Human Rights? In the case of anthropogenic climate change, addressing our global impact as a species means taking "differentiated responsibility" as nations. Sverker Sörlin and researchers from the Environmental Humanities Lab at the Swedish Royal Institute of Technology capture these tensions in a disarming question: "Where is the Anthropocene society?"

The Anthropocene is hardly a neutral scientific category, and it is being used to justify efforts of small groups to manage (and profit from) global earth systems. At another extreme are speculative works, such as Alan Weisman's *World Without Us*, that imagine how human infrastructure, wastes, and cultural monuments might fare in our absence, or Erik Conway and Naomi Oreskes' *The Collapse of Western Civilization*, which looks back on the twenty-first century from the bemused perspective of future Chinese historians who want to know how and why Western liberal democracies failed to tame rampant capitalism and mitigate global warming. These speculative visions imagine radical transformations, even as the worldwide recession after 2008 sharpened awareness of increasing social inequality. The "new era of capital" has provoked uprisings, from student protests in Montreal to the Occupy movement, the indignados in Spain, and economic unrest that fueled the "Arab spring."[31]

Yet not everyone sees the Anthropocene as a catastrophe. A few environmentalists have rallied to an optimistic futurism that has been dubbed the "good Anthropocene."[32] Stewart Brand, a futurist and the publisher of the original *Whole Earth Catalog*, amended his popular eco-philosophy to include a strong techno-scientific prescription: "we are as gods and have to get good at it." Andrew Revkin, a longtime environmental journalist for the *New York Times*, has projected a possible "good Anthropocene" in which societies take responsibility for anthropogenic climate change and other destructive actions and then act to mitigate them. Hopeful, yes—but is this scenario plausible?

Some humanists and social scientists who agree that humanity has become the primary driver of planetary biophysical and geophysical change nonetheless criticize the concept of the Anthropocene. It does seem to revive a Whig narrative of the Scientific Revolution and to invoke a belief in human ability to master nature. Yet in fact humanity struggles to understand and to predict, let alone to manage, the risks of complex systems. Reviving a discourse of Man's Control of Nature, Eileen Crist argues, is likely to justify further exploitation of ecosystems, animals, women, and marginalized social groups.[33] Andreas Malm and Alf Hornborg launched an equally broad critique of the Anthropocene discourse from the perspective of the social sciences. They suspect it cloaks a neo-imperial ideology that erases historical knowledge of the overwhelming agency of wealthy white Europeans. It was not all of "humanity" but a class of industrialists that extracted the coal and iron, built the factories, and set the average global temperature on its slow climb. The collective representation of "the human" masks the complexities of culture and history.[34] For similar reasons, few scholars in the global South have embraced the concept of the Anthropocene.

One test of the durability of this idea came when 30 of the 35 members of the Working Group on the Anthropocene recommended at the International Geological Congress in August 2016 that the term be adopted with a tentative start c. 1950. Other important tests occurred in public museums in Munich, Sydney, and Washington, in art galleries in Toulouse and Tokyo, and in lecture halls in Delhi and São Paulo. In interviews, very few potential visitors to the world's first major exhibit on the Anthropocene, held at the Deutsches Museum in Munich, had ever heard of the term.[35] Another measure of its usefulness may be whether this "boundary term" brings people into a new conversation that does not treat human agents as sovereign. As Dipesh Chakrabarty perceptively put it, humanity has "stumbled" into the Anthropocene.[36] Though natural scientists may have revived this conversation by inventing a new technical term, the philosophical and political questions that it has raised call for the expertise of the humanities.

Mining the Dark Mountain: Apocalyptic Narratives and Environmental Education

Narratives of the apocalypse are far older than Anthropocene discourse; they can be found in the Babylonian Epic of Gilgamesh and in the writings of Old Testament prophets. The genre was easily adapted to include cataclysmic climate change, atomic warfare, and genetic research gone disastrously wrong. In novels, films, and creative nonfiction, the end of the world became not an act of God (such as the great flood in the Old Testament) or the result of natural forces (such as a large meteor hitting the earth) but rather a result of human actions. Many people now inhabit places that resemble post-apocalyptic landscapes: slums, vast refugee camps, and regions laid waste by mining, or other anti-landscapes.[37] The specter of eco-apocalypse haunts the global North and is tied intimately to its rise to industrial power. The visionary English poet William Blake imagined a New Jerusalem on the other side of his hellish industrial present, the infernal mills of industrializing Great Britain. Later reformist and radical writers, from Zola to Shaw to Eliot, linked extractive industries to pollution and their power in a precarious and violent social order. Marxist theorists now write of a permanent "ecological rift" between capitalism and the earth and by turns predict and hope that future societies will organize themselves in radically different ways than by universalizing an "American way of life."

Rachel Carson opened her 1962 book *Silent Spring* with a terrifying yet familiar "fable" in which a pastoral landscape in the American heartland was poisoned by an invisible chemical dust. She appealed to public fears of atomic warfare that had been stoked by sensational journalism and radio dramas and by activist scientists who sought to inform the public of the dangers of nuclear technologies. A similar blend of expertise and emotive power animated scientific reports such as the Club of Rome's *Limits to Growth* and films such as *Soylent Green*. The post-apocalyptic genre blossomed during the era of mutually assured destruction and was a common trope in science fiction. It entered the mainstream after high-profile industrial accidents and mass poisonings such as Love Canal (an entire suburban community exposed to toxic sludge), Bhopal (3,800–15,000 dead from a chemical spill), Chernobyl (permanent evacuation of a city), and the BP blowout in the Gulf of Mexico (oil-soaked beaches and devastated fisheries). Nuclear accidents and large chemical spills cast a generational penumbra as young people adjusted to lives under a cloud.[38] The Asian Brown Cloud identified by scientists over the Indian Ocean in 2002 served as a further globalization of the landscapes of exposure.[39] As environmental risks widen, the culture industry of eco-doom and gloom prospers, as does the business of inventing and selling new technological fixes for environmental degradation. The spread of human-manufactured toxins and the inadvertent creation of new disease environments and vectors has created a new range of public health threats (and new opportunities for profit), such as avian flu in Indonesia and the West Nile virus. Many critics, including Ulrich Beck and Naomi Klein, have noted the capacity of present-day global capitalism to absorb and profit from a normalization of accidents.[40]

Many are disillusioned with environmentalism as it is practiced through the dominant institutions in the global North, such as the large NGOs, national protection agencies, and much of what goes under the headings "conservation," "nature protection," and "sustainable development." Some ecomodernists—or believers in eco-friendly modernization—are genuinely alarmed by climate change, but their solution is to double down on the civilization-via-capitalist-development wager.[41] They think that the problem of feeding 10 billion may be solved by satellite monitoring and more intensive technological transfer from developed countries to small farmers in Africa and Asia, and that site-specific remote sensing data combined with local soil sampling will give a more complete, higher-resolution picture of potential crop yields, down to the hundred-yard furrow. Techno-optimists evoke a "New Dark Age" threatened by extreme weather events in order to describe how it can be averted by a knowledge society.[42]

Poking holes in the bright balloon of techno-optimism is all too easy, however. Global agriculture has produced caloric surpluses for decades. The Food and Agriculture Organization reported in 2009 that the increase in hunger during the 2007–2008 food crisis coincided with a record harvest.[43] Hunger and malnutrition persist within otherwise overfed populations in rich nations, and primary rainforest has been cleared to produce beef for a dangerously overweight population in the global North.

The injustices of maldistribution have become impossible to ignore. However, anti-corporate violence and cynical hedonism are cheap intellectual and moral responses to environmental despair. More reasonable and practicable responses are calls to refocus our effort on the inhabited, human environment at the sub-planetary scale. For example, the historian and arts activist Jenny Price urges us to "stop saving the planet" and start doing the hard work of recovering urban rivers and neighborhoods.[44] Equally frustrated by status quo environmentalism, two British journalists and longtime environmental activists, Paul Kingsnorth and Dougald Hine, launched the Dark Mountain Project, which turns to the imaginative, creative realm of the arts, and particularly literature, to ground a vision for life in an era of darkening expectations. Their founding manifesto, titled "Uncivilisation," declared: "The Dark Mountain Project is a network of writers, artists and thinkers who have stopped believing the stories our civilisation tells itself.

We produce and seek out writing, art and culture rooted in place, time and nature."[45] The manifesto begins with a quotation from Ralph Waldo Emerson: "The end of the human race will be that it will eventually die of civilisation." The members of the Dark Mountain Project believe that "the world is entering an age of ecological collapse, material contraction and social and political unraveling." Twice a year they produce a book of 300 pages containing what they call "uncivilized" poetry and fiction.

The arts confront issues head on, and then refract this encounter with ecological crisis into plural visions and questions. How do we learn to turn toward rather than away from ecological degradation? How do we learn to sift out quick, marketable pseudo-solutions to find durable ideas and forms of coping? Is it time, as the biologist and writer Elin Kelsey has argued, for us to get beyond doom and gloom? Or is it appropriate to wear a kind of perpetual night in our souls as, one by one, entire forms of life disappear? Life in this twilight can seem particularly vulnerable, and care can seem a costly burden. How will we define our humanity when wading forward into a century of ecological loss? The arts can help us acknowledge and confront the melancholic, dark side of humanity's impact on the global environment.

Two fine examples come from the architect David Adjaye: a design for post-Katrina low cost housing in New Orleans and the Mass Extinction Memorial Observatory now under construction on the Isle of Portland on the southern English coast. These two pieces of effective place-making respond to the destructive ecological forces at loose in the world. Adjaye's post-Katrina housing is raised on stilts above potential floodwaters. Its design replaces the traumatic space of enclosed attics (many of which became tombs for trapped residents during the flood) with open third-floor decks that catch intermittent breezes during Louisiana's sultry summers. In contrast with deep environmental injustice, systemic corruption, poor levee maintenance, and the inept emergency responses after Hurricanes Katrina and Rita, such rebuilding represents both ecological adaptation and positive environmental justice.

Ecology and memory are equally joined in Adjaye's Mass Extinction Memorial Observatory (MEMO). E. O. Wilson helped to raise funds for it and broke ground to begin its construction in 2014. It is also supported by Sir David Attenborough, known for his many television programs on the environment, and James Lovelock, originator of the Gaia theory. MEMO is planned on the scale of a cathedral and will serve as a public space. It will house seminar rooms and a collection of sculptures representing lost species. A central oculus will focus light through the form of the portland spiral, a common shellfish fossil found in the local limestone being used for the construction. The memorial will both remember the more than 860 species that have gone extinct in our time and draw attention to the ongoing wave of extinction. A bell cast with fossil imprints by sculptor Sebastian Brooke will be rung in mourning of each extinction. Adjaye's work tries to capture the emotional content of its site as well as its physical form. The frozen upward sweep of the building's design dramatizes the complicated emotions that swirl around the ongoing biological losses human beings have caused. It evokes Pieter Brueghel the Elder's painting "The Tower of Babel" (1536) and also visually echoes the coastline's spiral fossil formations. The exposed outcroppings of England's "Jurassic coast" mark the site where the scientist Robert Hooke first wondered at the imprints of creatures plainly visible there in layers of Portland limestone. Hooke invented the English word "extinction" to describe the disappearance of forms of life. MEMO registers in its form both an upward, ascending energy associated with the aspiration of universal human knowledge (how impressive our catalog of life!) and a melancholic reflection on the partial, flawed nature of that knowledge and our seeming incapacity to coexist humbly with other earthbound creatures.

The Anthropocene has become an influential and contagious cultural idea—a meme—that evokes complex emotions: grief, fear, doubt, uncertainty, morbid curiosity, lethal rage. The critic Alexa Weik von Mossner, drawing on Ursula Heise's work on eco-cosmopolitanism and recent eco-criticism focusing on climate fictions, has argued that film and fiction might give audiences a better handle on the outsized emotions provoked by climate change, a key symptom of the new epoch.[46] Agnes Wooley, for example, has interpreted Jeff Nichols' film *Take Shelter* (2011) as a dramatization of the tension of internalizing and acting on climate awareness in a culture of denial.[47] As the protagonist dreams and experiences premonitions of a coming mega-storm, he begins to frantically dig and stock an emergency shelter for his family. Even the hero, Curtis, who accurately foretells a coming disaster, takes his anxiety as a sign of hereditary mental illness and seeks treatment. The film ends with Curtis with his wife and daughter on a beach and in a terrifying and ambiguous final shot, we see them staring at the ocean: a storm line with multiple cyclones approaches from the horizon. Is it a confirmation of a prophecy, the realization of a threat no one took seriously, or has Curtis's paranoia finally infected his family? Such films remind viewers of documentary footage of real environmental disasters, from Exxon Valdez to Deepwater Horizon, the Dust Bowl to Fukushima: the stock of traumatic imagery is vast.

Confronting Actual Disasters

The range of emotional responses to slower, multigenerational environmental disasters often surprises individuals caught up in them. It can be deeply disturbing to apprehend the way human tissues are permeable, inhabited, and vulnerable. What Rachel Carson diagnosed as biocide is ecocide and rebounds through bio-magnification up the food chain, from plankton to cellular toxicity in human bodies. The category of the human thus is endangered at the macro and micro levels. Memoirs by cancer survivors and by people affected by illnesses linked to pollution, such as Terry Tempest Williams' *Refuge*, are part of what the critic Lawrence Buell has dubbed "toxic discourse," the "expressed anxiety arising from perceived threat of environmental hazard due to chemical modification by human agency."[48] The emotional responsiveness and epistemological uncertainty of toxic discourse are as important as its political conclusions to regulate toxins and greedy corporations, an abiding element since Carson's *Silent Spring*. Likewise, atomic warfare and industrial nuclear disasters loomed large in international culture during the Cold War. Actual and imagined nuclear accidents provoked wonder, surprise, even a sense of disturbing beauty as well as sadness, anger, and a kind of pride in the self-sacrifice.

Svetlana Alexievich spent three years interviewing survivors of the Chernobyl accident—workers at the plant, the firemen (or "liquidators") sent in to clean up after the partial meltdown, doctors, teachers, scientists, refugees, and Party bureaucrats. "I felt like I was recording the future," Alexievich writes in her book *Voices from Chernobyl*. "These people had already seen what for everyone else is still unknown."[49] *Voices from Chernobyl*, which reads like dark prophecy, is an acute critique of the systematic lies that blanketed both the government response and the cultural memory of the event in Russia and the former Soviet republics. Many of the interviewees spoke with irony and fatalism; vodka was the self-medication and informal currency of choice, and many of their oral histories evoke a mythic, long-suffering "Russian soul." A hypothetical *cultural* trait of toiling through anguish and loss thus validates, retrospectively, a social-ecological disaster at the reactor. "We were told that we had to win," one liquidator remembers. "Against whom? The atom? Physics? The universe? Victory is not an event for us, but a process. Life is a struggle. An overcoming. That's why we have this love of floods and fires and other catastrophes. We need an opportunity to demonstrate our 'courage and heroism.'"[50] As a story-teller and a listener, Alexievich dares to look and record the moral evidence of the senses that contradicts the national trope of heroic, monumental sacrifice. A medical attendant holds out the medical cards for children born in 1986, many of whom have since died from thyroid cancers. "I hear about death so often that I don't even notice it anymore," says a literature teacher, Nina Konstantinovna. But her students "don't like the classics anymore." Why? The cultural canon does not speak to their experience as victims of planetary ecocide: "There's a different world around them now. They read fantasy books, this is fun for them, people leaving the earth, possessing cosmic time, different worlds."[51] Literary critics have argued that planetary-scale ecocide challenges conventions of representation (Nixon) or even the structure of narrative itself (Wiggin, Rigby). But perhaps the failure of representation also describes a generational gap, as people born into the Anthropocene era seek new forms to express and cope with their new reality.

Without a new vocabulary or convention for figuring post-disaster life, many revert uncomfortably to existing ways of life and deny or try to trivialize the disaster's lingering effects. Nina Konstantinovna describes with horror how she and her fellow Chernobylites continued their everyday lives. Peasants harvested radioactive vegetables from the fields: "Everything went on its way: they turned over the soil, planted, harvested. The unthinkable happened, but people lived as they'd lived. And cucumbers from their own garden were more important than Chernobyl. The kids were kept in school all summer, the soldiers washed it with a special powder, they took off a layer of soil around the school. And in the fall? In the fall they sent the students to gather the beet-roots. ... Chernobyl isn't as bad as leaving potatoes in the field."[52] Faced with an unprecedented, invisible exposure to radiation, many people simply continued to live according to their existing moral economy, in which wasting food is a more substantial sin than the invisible poisoning of children.

A filmmaker, Sergei Gurin, describes the beautiful spring of 1986 in Chernobyl as visually disorienting. The scale of the disaster relative to its low initial death toll and lack of spectacle challenged his skills as a war photographer sent to document the event. Finally, the biophysical impact of radiation on his body registered the disaster that could *not* be captured on film: "I'm holding the camera in my hands, but I don't understand it. This isn't right! The exposure is normal, the picture is pretty, but something's not right. And then it hits me: I don't smell anything. The garden is blooming, but there's no smell. I learned later on that sometimes the body reacts to high doses of radiation by blocking the function of certain organs."[53] Gurin's epiphany while photographing lilacs and apple trees in full bloom is typical of the alienated, surreal sensations reported by survivors of other large-scale environmental disasters.

Perhaps the greatest crime is assimilating the horror of preventable "accidents" into everyday life. *Voices from Chernobyl* offers damning evidence of how Soviet propaganda denied, discounted, and worked to erase evidence of the medical risks. "If the Swedes hadn't told," one resident farmer jokes, "we'd still be on our tractors, getting old." Yet the spread of disaster tourism and the accommodation of liberal democratic societies to predictable, ongoing ecocide offer no alternative. "Propagating risk and environmental crisis becomes," according to Frederick Buell, "the ultimate extension of consumerist capitalism, the creation of the kind of market that economists and industrialists would (perhaps literally) die for: a market that, in both theory and fact, can never, ever be glutted."[54]

There are also deep cultural and psychological impacts from large-scale modification of the earth's surface. Mining has created landscapes that appear otherworldly in their scale and devastation. In his series *Manufactured Landscapes* (2004–2007) the Canadian photographer Edward Burtynsky made mega-mines famous (or notorious) for their disturbing beauty. His lens captured the intense chemical greens, blues, and reds of tailing ponds, the blocky, childlike symmetries of pits seen from hundreds of feet above, and the puniness of the human body relative to the engineered scale of today's mines. It's the unanticipated beauty of sites that are profoundly degraded—dead, scarred landscapes—that produces the moral intensity of *Manufactured Landscapes*. Burtynsky arguably succeeds more effectively by sharing an uncanny vision than most who argue we must restrain our consumption. What is the meaning of this ongoing activity? Who is

responsible? Viewers are inevitably complicit. Burtynsky himself emphasizes his own entanglement with the material networks of trade, manufacture, and photographic equipment in Jennifer Baichwal's documentary on
the Chinese phase of the project.[55] *Manufactured Landscapes* come mediated
through screens made of refined oil and other minerals, perhaps destined
to end up in one of the epic middens of e-waste captured in his portraits of
China's e-recyclers.

Awareness of the manifold ecological costs of present-day life goes
under many names: petromelancholia, the ecological thought, slow violence, solastalgia. All point at the price to human and cultural meaning
that attends biological loss and planetary environmental change; they are
part of a critical vocabulary for the environmental humanities that heeds
Walter Benjamin's caution that every testament of civilization is also a
record of barbarity.[56] Timothy Morton has insisted we must tarry with the
melancholic, depressive side of what he calls the "ecological thought," his
term for the apprehension of our bodily, existential connections to systems
that plow away the earth's soils to sterile bedrock, clear its primal tropical forests, and fish down its ocean food chains. Mega-draglines, bucket
wheel loaders, and thousands of tons of explosives have transformed whole
regions of the globe, among them Australia's Hooker River, West Virginia's mountains, South Africa's gold mines, and Chile's copper fields. The
Australian psychologist Glenn Albrecht coined the term "solastalgia" to
describe the homesickness that can occur when one's home disappears
through radical environmental change. Albrecht and a team of researchers
have interviewed residents in New South Wales in areas affected by drought
and open-cut coal mining. They found symptoms of environmental physical and mental distress—what they call "psychoterratic" sickness—among
both aborigines and Australians of European descent.[57] The Louisiana poet
Martha Serpas explores a similar intense range of dark emotions in her collection *The Dirty Side of the Storm*. Among its other themes, the collection
takes up the loss of bayou land and culture to coastal subsidence, which
wetland biologists in the region have attributed to the canalization of the
bayou for oil and gas pipelines. She describes a now common misfortune in
the Anthropocene: "the steady vanishing / of your birthplace before your
eyes."[58] Stephanie Lemenager, reading Serpas among others, has developed
an extended analysis of Americans' troubled dependence on petroleum.
She diagnoses the American century with "petromelancholia," a state of

unhealthy attachment or "bad love" for oil that not only fuels culture and feeds filmic and literary representations but also infuses who Americans are and how they feel their way through the world.[59] False positive images are embedded in everyday phrases, such as "energetic activity," "electrifying performance," and "tanking up," which implicitly lay claim to a bonanza of once easily available energy. At the same time, more and more people find themselves dispossessed *in situ* of their country and lose the solace of familiar territory. Appraising the range of emotions, social relations, and cultural representations being produced after the Great Acceleration calls for artists and humanists to join the medical community in addressing what could be a global pandemic of psychoterratic disorders.[60] Critics and environmental humanists are not called to wish away ecosickness or hype the newest pill to purge eco-melancholy. Rather, the arts and the humanities explore insights, often marked by experiences of illness, that reconnect bodies and minds to the earth.

Yet human beings do not suffer the greatest harms. The ecological effect of radical strip mining is permanent for endemic flora and fauna, who have nowhere to go once their habitats are destroyed. Deep forest vegetation communities depend on soils that take hundreds if not thousands of years to develop; post-mining reclamation focuses on approximating land grade and profile and limiting soil erosion—not on wholesale restoration. In the industry's parlance, ecology is *overburden*, a euphemism that discounts the value of every niche and creature in the thin slice of living soil and tree canopy that was once above the extracted minerals.

In *Lost Mountain*, Erik Reece describes the insidious effect of such language on perception of mining's costs to the southern Appalachian region. He follows a biologist, Jim Krupa, and his team of students who are tagging threatened flying squirrels (*Glaucomys volans*) in Eastern Kentucky's Robinson Forest. With 12,000 acres of contiguous forest, Robinson is a vital habitat for North America's only gliding mammal, a smaller relative of the common gray squirrel that relies on mature forests with plentiful acorns, beechnuts, and hickory nuts as well as older trees with hollows where they can evade predators. Their population has declined as a result of forest fragmentation, hastened in Appalachia by mountaintop removal mining. Reece describes his work with Krupa: "We caught ten flying squirrels that July day in Robinson Forest. Their flights were all different and remarkable. As we stuffed the final cages in backpacks and walked back to Krupa's truck, I asked for his prognosis of the human condition. 'Oh, I think we're doomed,' he offered cheerfully. 'With our levels of population and rates of consumption, it's just a matter of time before we kill ourselves off.' He paused, wiped his glasses on his T-shirt, and smiled. "It's not something I tell my freshmen."[61] Krupa's response echoes what Aldo Leopold said about the effect of an ecological education. Many now see the "world of wounds" when they trek through old growth forest, dive on a coral reef, harvest olives on a traditional terraced farm, or view islands disappear under rising tides.

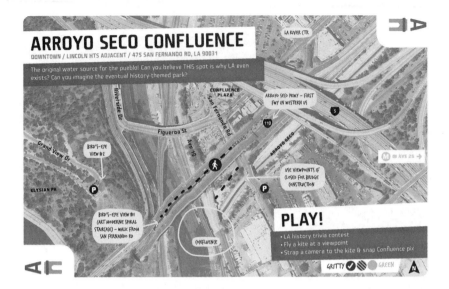

Figure 6.1

Art actions bring new people together to imagine alternative futures for existing environments, often in unlikely places. Black and white of original full color playing card, "Arroyo Seco Confluence," from the *Play the LA River* card deck, © 2014 Project 51.

6   Putting the Brakes On: Alternative Practices

Non-Apocalyptic Alternatives

The environmentalist David Suzuki once wrote: "We're all in a great big car driving at a brick wall at 100 mph and everybody is arguing over where they want to sit. My point is it doesn't matter who's driving. Somebody has got to say, 'For God's sake, put the brakes on and turn the wheel.'"[1] Suzuki has not despaired and believes we can change course and avoid self-destruction. This chapter will explore alternatives to the dark visions described in the previous one, including localization, degrowth, advanced recycling, and commodity regionalism—all efforts to redefine economics so that it takes account of the environment. These ideas and movements can be classified as part of global environmentalism, a decentralized movement that has found expression on every continent. Paul Hawken estimated in 2007 that there were 2 million organizations involved, with many different orientations, ranging from NGOs to groups protecting a particular forest or wetland area to those fighting against the introduction of genetically modified foods to indigenous peoples resisting misuse of land leases to local food cooperatives. Hawken has surveyed and interpreted this heterogeneous development in *Blessed Unrest*, which captures its scale and pluralism.[2] He sees it as "a global humanitarian movement arising from the bottom up" that may be the largest social movement in the world's history. It is stimulated to action by a common realization that "the planet has a life threatening disease, marked by massive ecological degradation." The grassroots are responding to this realization, and they are aggregating into networks, locally and internationally. Hawken hopes that this process will lead to a "conspiracy of social imaginaries" that will cultivate and share knowledge that leads to a groundswell of opposition and social transformation.

His ameliorist vision points to a peaceful transition to a more balanced global system. He notes that the world's largest 200 corporations have more wealth than "80 percent of the world's people, and their asset base is growing 50 times faster than the income of the world's majority." Yet his book is not a call to the barricades, and Hawken himself is a successful entrepreneur. In *The Ecology of Commerce,* he urges businesses to embrace environmental values.[3] What he means by this is based on biology, which shows that life, starting at the cellular level, builds from the bottom up, assembles itself into chains, and generates variations. This means to Hawken that grassroots groups "are the most efficient social entities on earth," as demonstrated by microlending in Bangladesh, non-profit health care in Haiti, or farming based on local knowledge.[4] It follows that Hawken and millions of others believe there are alternatives to the apocalyptic visions explored in the previous chapter. One is "localization." We touched on this idea in chapter 1's discussion of the commons and in chapter 2's discussion of the centrality of recovering a sense of place.

Localization

The focus on place has been by no means limited to cultural geographers, essayists such as Wendell Berry, or literary critics such as Lawrence Buell.[5] Many in the social and natural sciences have taken a similar tack. Two professors of natural resources and the environment at the University of Michigan, Raymond De Young and Thomas Princen, argue that the environmental damage caused by the high-consumption global economy, coupled with emerging shortages of resources, makes a transition to local economies unavoidable.[6] Globalization has been centrifugal, based on inexpensive energy and abundant raw materials, and it has reinforced the centralization of political and economic power. In contrast, localization is centripetal, decentralizing power into regional communities and using resources sustainably. Globalization is based on an ideology of continual growth and treats waste and pollution as externalities that at times are exported to Third World dumping grounds. In contrast, localization is based on sustainable growth that includes recycling and improvements in design and manufacturing, in order to minimize waste. People can reduce resource use and yet improve their quality of life. Localization is a logical response to the Anthropocene, based on the realization that less wasteful ways of living are necessary if anything like the ecological system as we know it is to survive. Localization is not a movement based on apocalyptic fear. Rather, De Young and Princen argue, "People will intuitively see that localization can be a force for good."[7] This transition will not be easy, however, and they argue that it will demand an escape from "top-down, elite-driven global management."[8]

Why not rely on experts to find the best solutions and impose them quickly from the top? Because idealized solutions ignore the complexity of the world. The World Bank and many other organizations have learned through bitter experience that what seems a perfect way to promote reforestation or to improve irrigation in one location is not always transferable to another site. Local ecologies, geologies, and cultures are different. What works in Sweden or Italy cannot simply be imposed in Manchuria or Kenya. There was a period when enormous dams were built in many parts of the world, based on the successes of the American Tennessee Valley Authority, which during the 1930s constructed 21 dams that together prevented flooding, improved transportation, generated electricity, and accelerated the "modernization" of an economically depressed region. This seemed a model worth emulating globally. However, once many such projects were built, "in southern countries the large artificial lakes often brought an ecological fiasco in their wake, whether through enormous silt deposits, the increased evaporation, or the breeding grounds of epidemic diseases provided by the standing bodies of water." Many of the dams eventually were understood as "disastrous mega-technology."[9] Not every dam was a mistake, but the idea that one solution fit everywhere was seriously mistaken. Similarly, seeds well suited to Peru may not grow comparably well in Jordan or Indonesia, and tractors used in large flat tracts of well-watered Illinois farmland may proved ill-suited to dry, hilly land in Africa or Australia. One needs to know the local ecology and how the local agricultural system works with that ecology.

At the other extreme from international aid organizations are the many ecovillages that have sprung up around the world in the last half-century. Unlike top-down institutions, they emerged without much fanfare. In 1991 they held an international meeting that led to the creation of the Global Ecovillage Network (GEN) in 1994, including a website.[10] There is no screening or evaluation before membership is granted, and the communities are diverse. Many are in Sri Lanka, Senegal, Latin America, and other non-Western countries. Some blend ecological reform with a spiritual dimension, notably the Sarvodaya project that includes 15,000 villages in Sri Lanka's largest non-government organization. Most ecovillages are separate, however, with loose connections to like-minded communities. While they vary considerably, all seek to maximize self-sufficiency, democratic participation, recycling, and education. As Karen Litfin summarizes, "Beneath this commitment to social and ecological sustainability, one may discern a worldview premised on holism and radical interdependence" that is fundamentally different "from the assumptions underlying modern consumerism." These are not protest movements that focus on trying to reform existing structures, though individual ecovillages many have links to groups struggling for agrarian reforms, such as Brazil's Landless Worker's Movement (MST). Rather, they are "creating parallel structures for self-government in the midst of the prevailing social order."[11] They demonstrate that sustainability is not just an idea or a luxury for the middle class, but an achievable goal. Litfin traces the emergence of this movement to a variety of sources, including the Gandhian movement and Schumacher's *Small Is Beautiful*, but she singles out for emphasis a holistic approach developed in Australia in the 1970s called "permaculture" that has been taken up and modified in many ecovillages. Its central principles are to design based on nature, to capture and store as much alternative energy as possible, to intervene in the local ecology as little as possible, to adopt slow solutions to problems, to minimize and recycle waste, and to value diversity in both human culture and agriculture. Bill Mollison and other gurus of permaculture have mastered the arts of layering agricultural production in "food forests" that mimic forest ecologies, but produce an astounding variety of foods in dense patches. Mollison's own work, growing a food forest on a desiccated former sheep station in Australia, dramatizes the restorative power of agro-ecology. The object of permaculture is to escape from the fossil-fuel economy of consumption and waste into a sustainable local economy. The practice of permaculture has spread to all parts of the world, with more than 2,000 projects, for example in Ecuador, Brazil, Portugal, Britain, Sweden, Latvia, Indonesia, India, Mexico, Mongolia, Italy, Canada, and the United States.

It might seem logical to suppose that ecovillages and permaculture projects are inward-looking communities that are regressing to pre-industrial forms of living, but they are acutely aware of world developments, and hope their movement can provide a blueprint for future living. They are "an affirmation movement, not a protest movement" and they "tend to be active in local, national, and transnational politics."[12] They oppose globalization policies that ship jobs around the world, but they are not isolationists. Ecovillages in Denmark sent thousands of bicycles to Senegal, and a Portuguese ecovillage engaged in conflict resolution in Columbia.[13] They also have established regional education centers to help spread their knowledge of organic gardening, water recycling, alternative energy, and building from local materials such as straw bales. The ecovillage movement is by intent not centralized, and there are few statistics available to measure its growth or decline. But it appears that the most successful enterprises are those that have a strong religious dimension, as is the case in Sri Lanka, or a potent political vision, as in Latin America. This was true of nineteenth-century utopian communities as well.[14] Those that were secular seldom lasted even a decade, while religious groups like the Mennonites or the Amana Community in Iowa, persisted.

At an intermediate level between ecovillages and transnational organizations, some cities have decided to reduce their environmental impact. This trend was briefly discussed at the end of chapter 2, but to exemplify how this works in practice, consider the Danish city of Sønderborg. With a population of 77,000 people, Sønderborg committed itself in 2007 to becoming carbon neutral. Its "Project Zero—Bright Green Business" set a deadline of 2029 to become carbon-free. It did not act alone but coordinated its efforts with the World Wildlife Fund, the Chinese Academy of Social Science, and the University of Southern Denmark. As part of a consortium of cities with similar goals, it helped to develop a Low Carbon City development Index (LCCI) to create a common standard to measure progress. Sønderborg began to abandon fossil fuels in favor of solar power, offshore windmills, geothermal energy, ocean-water cooling, and burning biogas from the waste of local pig farms. New houses in town must be energy neutral or even net producers of electricity. The commitment to this vision attracted more than fifty clean energy-related businesses and startups to the city. As part of the plan, Sønderborg is concentrating new construction and growth along a south-facing waterfront area, increasing its population density, making bicycling more attractive, improving mass transit, and reducing car traffic.[15] Sønderborg is one of many cities seeking to become the world's first with zero-carbon emissions. Another small Danish city, Skive, shares the same goal. Two Australian cities, Melbourne and Adelaide, have even more ambitious deadlines.

However, residents of the Danish island of Samsø are not much impressed. They have few automobiles but many bicycles. They once imported all of their energy in the form of coal, oil, and gas, but between 1997 and 2007 they converted to 100 percent renewables, primarily wind power plus some solar and burning of biomass. Samsø now has a surplus of electricity and exports it to the mainland. The next goal is to become completely free of fossil fuels for any purpose whatsoever. Samsø and the mainland town of Skive together offer tourists an "energy safari" to see how their plans are being implemented.[16] Samsø is fast becoming quite literally what David Hess calls "an energy island," a self-sufficient locality,[17] and this model seems to have inspired Dutch and Danish planners and designers. In 2016 they were developing plans for self-sufficient communities that include agricultural production, located in or near large cities. Their energy efficient houses will have no electricity bills, and they will recycle gray water into greenhouses to provide fresh, ecological vegetables. There will be fish breeding pools and poultry production, which demand less energy to produce than red meat. The first such community is being built outside Amsterdam, in Almere, a new city of 200,000 built on land reclaimed from the Zuider Zee, which will also host the 2022 World Horticultural Expo (Floriade). Almere is part of a group of instant cities linked in the International New Town Institute. Each has something to teach the others in the network. Alamar, Cuba has highly developed urban gardens; Curitiba, Brazil, runs a highly successful recycling program where residents can exchange 4 kilograms of organic waste for one kilo of fresh vegetables; Chandigarh, India shows how to adapt classic modernist planning by Le Corbusier to the environmental needs of the future.

The international "Transition Towns" movement has also spread worldwide since 2005. The number of "Transition Towns" expanded rapidly after the economic crisis of 2008. One pillar of the "transition" movement is that citizens should prepare now for climate change, economic crisis, and an inevitable lower energy future by re-skilling and strengthening their social resilience. These initiatives operate around the globe. They are organized autonomously but linked through online discussion platforms. Volunteer Transition Town leaders organize "re-skilling" workshops to educate neighbors in beekeeping, veggie diesel conversion, aquaculture, and many other crafts of self-sustenance. Perusing eco-magazines (e.g., *The Permaculture Activist*) and independent media websites reveals much overlap between the Transition movement, permaculture, urban community gardens, and agrarian movements like Via Campesina, all of which envision a bright but lower energy, lower carbon future.

Degrowth

The island of Samsø illustrates a successful conversion to zero-carbon emissions. An important movement has also developed around the concept of "degrowth," a term first used in scientific fields such as aquatic biology but later adopted as a central concept in an alternative economics paradigm. The term can be traced to the writings of Nicholas Georgescu-Roegen, who became prominent in the early 1970s after publication of *The Entropy Law and the Economic Process* (1971). While briefly associated with the Club of Rome, he rejected its focus on sustainable growth and instead argued that economics ultimately had to be grounded in ecology, which made degrowth inescapable. This argument was scarcely audible during the neoliberalism of the 1980s and the 1990s. But degrowth regained influence as the crises of global warming and species extinction worsened, and it was one of the ideas that underlay a 2002 UNESCO symposium, "Unmaking development and remaking the world."[18] There are regular conferences on degrowth. The University of Barcelona hosted the 2010 conference, with 500 people attending from 40 countries. Similar events were held in Montreal and Venice in 2012 and Leipzig in 2014.[19]

The movement is developing ways to shrink the economy. This will demand a shift away from mining and smelting to recycling, and it will end the focus on consumerism as a central driver of the economy. Products will be designed to last rather than becoming quickly obsolescent. People will buy less and share more. One group of academics has formed Can Decreix, a community that will pursue degrowth both as a way of life and an object of study. They note, "voluntary simplicity is not a goal in itself" but that "the search for simple, energy-saving technologies is intended to raise awareness of alternatives to overabundance, and to create social leeway."[20]

Returning to the Kaya identity discussed in chapter 3, degrowth can be achieved in many ways: by reducing consumption, minimizing waste, improving carbon intensity, increasing carbon efficiency, reducing the birthrate, or some combination of all of these. Under degrowth, some contraction in the GDP could be achieved by buying more locally sourced food.

In Montreal, for example, Lufa Farms operates two large greenhouses on rooftops and supplies 4,000 customers.[21] Driving more energy-efficient cars, people could still travel as much as before, but use half the energy. Some degrowth could also be achieved by taxing high incomes, as in the 1950s, with the goal of discouraging excessive consumption. The so-called "free market" has been socially constructed to generate unsustainable growth, and it needs to be reconstructed to foster sustainable degrowth. Rather than stimulate consumption through advertising, it would be dampened by restrictions on advertising. The quality of life for most people would not necessarily decline. People would work fewer hours; human and ecological health would likely improve. Advocates describe lives of dignity, with being placed ahead of having.

Degrowth should be sharply distinguished from what Peter Dauvergne has called the "globalization of environmental management," or the export of environmental regulations by organizations such as the World Bank and the International Monetary Fund. While in theory a good thing, all too often in practice such management imposes "an unbalanced process of economic globalization that draws down natural resources and deflects the costs of rising consumption away from those who benefit most and toward those who benefit least."[22] London and Los Angeles benefit from the new environmental regulations, but African and Asian ecologies often do not. Why? Because, as the world economy tripled in size between 1970 and 2000, Western economies proved adept at deflecting waste and pollution to emerging economies. At the same time, debt restructuring in Africa, Latin America, and the Caribbean often came with the obligation to develop raw materials—often forest and mineral wealth—for export. For example, during these years Japan's forests were carefully tended. It became one of the most heavily forested countries in the world, even as Japanese industry imported millions of less expensive logs from Indonesia, the Philippines, and Malaysia—countries that experienced deforestation.[23] The high labor costs and environmental regulations in Japan meant that its consumption of less expensive imported wood cast a long shadow over the rest of Asia. Similarly, British, French or German imports of raw materials undermine biodiversity elsewhere. Likewise, pollution caused during manufacture of electronic goods remains in China, while the mobile phones and laptops are shipped to Europe and the United States. In short, the world appeared to be embracing "sustainability" and "environmental management," which meant some improvements in efficiency, recycling, energy use, and the like, all while shifting the burdens of pollution and streamlining extraction. But from a global perspective such growth was unsustainable.

Some advocates of degrowth believe that their movement demands civil disobedience, to protest destructive practices. In August 2015, for example, 1500 people occupied an open pit coal mine in Germany. They were protesting not only the global warming that comes from burning coal but also the ideology of extractivism. The Tagebau Garzweiler mine in the Ruhr region is 12 kilometers wide and 100 meters deep, and every day 2,400 railway cars carry away more coal, as one loaded car departs every forty seconds. The protest stopped the excavations for just one day, but they drew attention to the German reliance on coal and the close alliance between the mine and the state.[24] Hundreds of police showered the peaceful protesters with pepper spray, as many people saw on the evening news. In May 2016, European environmentalists converged again on a German brown coalfield in the Ende Gelände action. More than a thousand activists blocked rail lines to the Welzow mine and demanded an end to burning coal for electricity. Such confrontations are symbolically powerful; they call attention to the intransigence of governments and world leaders in the face of climate change. At the same time, many in the degrowth movement focus less on trying to change the existing infrastructure than on creating off-grid alternatives where people can develop new, low-consumption lifestyles.

In a sense, degrowth has been taking place since the early 1970s, so far as ordinary people are concerned. Wages for 95 percent of the population in Britain, the United States, and many large economies have stagnated for decades. Efficiency increases once led to higher wages for workers, but increasingly managers and stockholders have profited while workers have not. During the 1970s the typical American executive salary was 42 times the average employee's wages. But thirty years later, as the economist Joseph Stiglitz notes, "CEOs were getting more than 500 times the wages of the average employee."[25] In fact, after adjusting for inflation, average workers scarcely made more in 2016 than their father or mother had in 1970.[26] The French economist Thomas Piketty has documented in detail this shift in wealth and the erosion of the middle class, a process that has been occurring for decades. Piketty notes one obvious solution to the problem: the reintroduction of progressive taxation, including a direct tax on wealth. This, combined with higher wages, would establish a fairer distribution of the benefits from increased efficiency. But a return to the income distribution of c. 1950 would likely stimulate unsustainable growth. Instead, much of this money is needed to (1) escape from the fossil-fuel economy, by increasing energy efficiency and shifting to alternative energies, (2) promote birth control and public education to encourage degrowth of the population, (3) create "green" cities that pollute less, (4) convert agriculture to "greener" practices, and (5) implement comprehensive recycling programs to reduce resource depletion. In short, the priority should not be simple wealth redistribution but a redesign of the economy for a sustainable society.

From Cradle to Cradle

One recurrent good idea for reducing human impact on the environment is recycling. During World War II governments discovered how much could be recovered this way, and today the challenge is to motivate the public to embrace the practice. In *Natural Capitalism* Hunter and Amory Lovins teamed up with Paul Hawken to inspire corporations to see environmentalism not as a foe but as a business opportunity.[27] They described, for example, how an Atlanta floor covering company doubled revenues and tripled profits through recycling. When such practices are combined with retrofitting buildings to use less energy, the result is growth while using fewer resources. The architect William McDonough and the chemist Michael Braungart did exactly that for the Ford Motor Company as they retrofitted parts of its River Rouge factory so that the water discharged from the plant was cleaner than before and the energy use lower. McDonough and Braungart subsequently became leaders of a movement to improve recycling. In their "cradle to cradle" systems automobiles and other complex products are designed from the outset with the goal of recovering every part for reuse.[28] Older recycling systems shredded entire automobiles, creating a mixture of high-grade steel, copper, and other metals. When smelted, these could not be used to make new car bodies or other high-end products. In theory "cradle to cradle" systems avoid such "down-cycling," as materials are separated and recovered in pure form. The idea is to create a "technical metabolism" where little or nothing is lost in the circulation of materials through systems. This model of industrial production is also sometimes referred to as "biomimetic design" or "regenerative design."

These are not mere speculations, as the European Union passed an "End of Life Vehicle Directive" in September 2000. It makes automobile manufacturers responsible for designing cars that can be disassembled for resource recovery. If and when such practices are extended to all consumer products, it will decrease solid waste, reduce the demand for new raw materials, and shorten the (re)supply chain. It is an effective degrowth practice. To effect this transformation, the Cradle-to-Cradle Products Innovation Institute (San Francisco and Venlo, the Netherlands) has a certification program for manufacturers, publishes guidelines for recycling, and advises consumers on which products meet its standards. Among the more than 3,000 approved products are house shingles, wallpaper, flooring, detergents, carpeting, and paint. There are Cradle-to-Cradle conventions where manufacturers exchange ideas and best practices.

By definition, these activities require interdisciplinary cooperation between architects, product designers, production engineers, marketing departments, and managers. Practices like biomimetic design suggest that if all products were thoroughly recycled it would focus economies more at a regional than a national level. Vast flows of raw materials would no longer be necessary, as each region could recover much of the copper, iron, tin, steel, plastic, wood, and other substances it required. The energy once expended to ship these raw materials could be saved, and tracking internal material flows would become a normal part of regional planning. This innovative idea is still in the process of being converted into practice. There are clear successes, such as the use of recycled plastic to make airplane seats, but a much-ballyhooed "green city" project in China (described in chapter 1) was stalled and came to nothing after local residents resisted it. Nevertheless, *Cradle to Cradle*'s concept of creating a technical metabolism remains inspirational, setting a standard for the more effective recycling needed to realize either sustainable growth or degrowth. McDonough and Braungart in *The Upcycle: Beyond Sustainability—Designing for Abundance* take these ideas further, arguing that the effect of manufacturing need not be environmental degradation.[29] The water coming out of a factory can be purer than that which went in. In short, it is not enough merely to do no harm. There could be an improvement whenever human beings intervene.

Likewise, as McDonough and Braungart's work makes clear, the arts are not merely a veneer of style to be added near the end of product design. Rather, the arts are fundamental to the definition of cultural styles and identities. Poor environmental behavior sometimes results from a lack of coordination, a failure to see individual buildings and products as part of an entire system that shapes identity and consciousness. The fine arts are best understood neither as mere decoration of surfaces nor as comic relief nor as beautification projects, but as ways to reconceive the place of human beings in their ecologies. For example, in San Jose, Andrea Polli designed an interactive sign "Particle Falls" that looks like a waterfall on the corner of a building, one that changes its appearance depending on how much particulate matter is in the air, making visible otherwise invisible pollution.[30] Designers have also created interactive monitors that show people exactly how much electricity they use relative to their neighbors, encouraging healthy competition to reduce energy use. Again, the invisible is visualized, encouraging different behavior. These are relatively simple but effective interventions. At another level, architects, engineers and designers working together can create entire environments—a house, a factory, or shopping street—that use less energy than a conventional facility, combined with ergonomic, modular furniture from upcycled materials, extensive use of natural light, and improved air quality.

The buildings of the future could resemble the Center for Sustainable Landscapes (CSL) at the Phipps Conservatory and Botanical Gardens in Pittsburgh.[31] It was built on a brownfields site, and in the spirit of cradle to cradle used as many recycled materials as possible. It was designed with the goal of being the world's "greenest" building. The CSL generates all of its energy, primarily from solar panels and wind turbines, and it uses only one-third as much energy as a conventional building. About 70 percent of its heat and cooling come from a geothermal system that penetrates 500 feet below the ground. The complex treats its storm and sanitary water and recycles it on site. A high-tech facility, the CSL also embodies "biophilia," a concept developed by Erich Fromm, taken up by the biologist E. O. Wilson, and defined as "the innately emotional affiliation of human beings to other living organisms." The CSL developed the project Biophilia Enhanced Through Art (BETA) with paintings, photographs, and art works throughout the complex.

Commodity Regionalism and Environmental Art

The idea of localization and the increasing focus on improved recycling and design can be linked to "commodity regionalism." This mode of qualitative critique has found expression in the works of Jenny Price, Mike Davis, Alan Sekula, William Cronon, and Richard White, and may be understood as a narrative elaboration of commodity chain analysis in sociology and political economy. The idea is straightforward enough, though carrying it out is challenging: to trace the origins, processing, and geographical trajectory of a product, or a cluster of materials needed to make a product. One practitioner sees "the transnational as the most fundamental if elusive space of economic globalization" that "tends to be most visible in regional sites of capital production and transshipment."[32] A suit may be made from wool sheared from New Zealand sheep, spun into thread and made into fabric in India, cut and sewn in Eastern Europe, furnished with labels and buttons from Britain, and then sold throughout the EU. Tracing such movements has become easier through the development of databases such as the Land Use Database developed in Los Angeles. Stephanie Lemenager, for example, traced the cultural trajectory of petroleum from well to user, including an array of sensory experiences (notably films, such as *Giant*) in which oil becomes a commodity and intimate extension of our lives from lip balm to garden hoses, plastic bags and dental polymers. Studies of this kind increase awareness of how wasteful and seductive our relationship with oil has become, not only in the obvious burning of it to run automobiles, but also in the destruction of habitat and in patterns of cultural production.

The advantages of biomimetic design become clearer when considered alongside the material flows embedded in globalization. This critique is akin to Martin Heidegger's discussion of how modern society tends to treat nature as a "standing reserve,"[33] which is echoed in Stacy Alaimo's comment that "rather than approach the world as a warehouse of inert things we wish to pile up for later use, we must hold ourselves accountable to a materiality that is never merely an external, blank, or inert space."[34] The living ecology of the more than human world represents, in other words, a richer materiality than that of the commodities resulting from our networks for appropriating, processing, and consuming nature.

Many late twentieth-century artists associated with the Land Art movement foregrounded precisely this tension between commodification and materiality of environments. Individual artists associated with "Land Art" held varying ideas of the environmental dimension of their work. Some, including the British land artists Richard Long, Chris Drury, and Andrew Goldsworthy, sought self-consciously to change the quality of human connection to nature as a primary objective of their work.[35] Others (among them the well-known American Robert Smithson) seem to have been largely indifferent to the ecology where they worked. And the site-specific sculptures, terraforming, and performances of Land Art took radically different forms: from Long's 1968 "A Ten Mile Walk, England," in which the artist walked a straight 10 mile long line across rolling moors; to Agnes Denes's planting of a wheatfield on the Battery Park Landfill, a block from Wall St. ("Wheatfield: A Confrontation," 1982); to Goldsworthy's ephemeral sculptures of leaves, twigs, thorns, or snow, designed to be observed as they melt and decay. Yet such works undoubtedly engage viewers in different ways than museum encounters and inspired later generations to expand how we appreciate, sense, and feel nature. Present-day environmental art activism engages with a similar impulse. Project 51's "Play the LA River" is a pack of illustrated playing cards. Each card guides visitors to rediscover one of more than fifty sites along the mostly forgotten, highly engineered Los Angeles River, inviting participants to sketch landscapes, write poems, and tell the hidden history of the city's founding at a watery confluence. But the LA River is best known as the concrete-jacketed setting for car chases in *Terminator 2* and *Fear the Walking Dead*. It is nature dead, buried, and repackaged for a deluxe box set entertainment. Environmental art like "Play the LA River" re-humanizes the alienating topography of this intensely controlled landscape to ask how it might instead link the urban region in more lively ways.

The problem of commodification, at another level, is much like that of inventing wilderness: one must avoid imagining humanity and nature as existing in separate realms. The food on the supermarket shelves and the goods in the warehouse came from somewhere, but people tend to forget the material flows. With the invention of barcodes on containers, these flows are now easier to track, and at least in theory enable researchers to follow a commodity across the globe as easily as they follow a package shipped from Fedex. This in turn would make it possible to construct a commodity's environmental footprint. Indeed, photographer Alan Sekula created *Fish Story*, a hybrid work of text and photography that documents the transformation of harbors and shipping after containerization, based on visits to many ports around the world as well as taking passage on a container ship.[36] Sekula deliberately crosses many disciplinary boundaries as he situates his work in the histories of painting the sea and of documentary photography. Among the many purposes of this assemblage was to re-materialize the abstraction of globalization, reconnecting it to visible processes, specific places, and identifiable people.

Ecological Economics

The ecovillage and degrowth movements, the many cities converting to "green energy," and the academic community have reached a consensus that climate change is real and that human beings must make fundamental changes in their patterns of consumption and organization of society. However, many politicians and voters have reacted like patients who go into denial when told that they have a life-threatening illness. In the United States the Republican Party went into denial about global warming. President George W. Bush tried to silence government scientists whose research proved that climate change was not just a theory. Though one would have hoped political leaders in 2017 would be more enlightened than George W. Bush in 2001, some still were in denial. Among the most powerful were those most mired in the political and economic systems that created the problem. They still conceived economic growth as the panacea for human problems such as unemployment, and translated all growth into the apparently neutral language of numbers. In this logic, the production and sale of 1 million gasoline-powered automobiles (along with the building and repair of roads for them to use) was just as good for society as spending the same amount to construct mass transit. By extension, the sale of electric cars for $1 billion dollars has no more environmental value than a sale of conventional cars for the same amount. Producing a million plastic bags would seem to have the same worth as producing sturdy reusable bags of biodegradable materials. From the perspective of conventional economics, even a disastrous oil spill is a good thing: it stimulates the economy, creating jobs to clean it up. This example is hardly theoretical. The British writer John Lanchester has sourced the creation of the "credit default swap"—a notorious financial product at the heart of the 2008 global economic crisis—to J. P. Morgan's innovative efforts to extend credit to ExxonMobil while avoiding tying up too much of its own capital in required reserves. Why? ExxonMobil needed $5 billion to cover damages paid out after the 1989 *Exxon Valdez* spill.[37]

As these examples suggest, the accounting methods of traditional capitalism are not designed to take nature into account. Efforts to design a replacement for gross domestic product (GDP), the classic growth indicator, began in the 1970s during the energy crisis. There is no agreed upon standard measure, however, though the World Bank has promoted the use of an indicator called Genuine Savings (GS). The problems in creating a new, "green" standard are complex, because they ask one to place a monetary value on such things as the work done by a forest in capturing CO_2 or preventing flooding.[38] An accurate new accounting system would track an "eco domestic product" (EDP) that, at a minimum would subtract from GDP (1) the depreciation of infrastructure and (2) depreciation of natural resources caused by economic activity.[39] Knowing the EDP, and whether it is rising or falling, is essential in order to develop a model of sustainable national income. Were EDP used instead of GDP, some projects would no longer appear sensible. By including environmental depreciation as part of the calculus of decision making, bus systems or light rail could well become decisively less expensive than building a new highway. New natural gas pipelines might become vastly more expensive than fully automated battery factories and massive solar or wind installations. No society is likely to adopt EDP without a debate, and environmental humanists will be needed to explain the alternatives. Convincing corporations and stock exchanges to accept a new standard such as EDP will call for new narratives of social well-being. The work of shifting from old models of social progress to new ones will find scholars in the environmental humanities making common cause with sociologists, political scientists, and economists. Important shifts have occurred in these fields in recent decades. Ecological economists have gained a professional footing in some countries, and research on alternatives to endless growth has flowered, particularly in Europe and Latin America. The abbreviation EDP has not caught on, however. Rather, "green accounting" has gained some favor, particularly in Germany.[40]

On a popular level, the most common term has become "ecological footprint," or the narrower "carbon footprint" which refers only to CO_2. During the 1990s considerable effort went into calculating the ecological footprint of 52 major industrial countries. More than half were using more resources than they had. This meant that Germany, for example, despite many "green" initiatives, had a footprint of 5.3 hectares per capita, but only a biocapacity of 1.9. The United States had a footprint of 10.3 but a biocapacity of 6.3. The world as a whole, it was estimated in 1999, was devouring resources as if the planet had 30 percent greater capacity than it actually does.[41] The "ecological footprint" is a vivid image, and the numerical conclusions seem easy to understand, but, as Philip Lawn explains, these statistics do not measure human well being or such matters as species extinction.

Taking a longer historical view, Bill McKibben has argued that the idea of growth as the goal of an economy is a misguided inheritance from the eighteenth century. He shows why continually wanting more is no longer a plausible goal, while making an effective practical appeal in his book *Deep Economy*.[42] It reprises many ideas of localization, sustainable cities, recycling and degrowth, and makes a case for community-centered agriculture instead of industrial scale agriculture, with its intensive use of oil and chemicals. Like the degrowth movement, he rejects globalization as a means to solve environmental issues. What is lacking are widely accepted indicators used by governments and financial institutions. However, they still cling to the narrative of progress that is reified in the concept of the Gross Domestic Product.

On a purely technical level, the problems presented at the beginning of this chapter, of a world economy accelerating like a speeding automobile toward a brick wall of resource depletion, global warming, species extinction, and overpopulation, can be solved. It is possible to put on the brakes. Degrowth is achievable in mining and extraction industries and it might be accomplished without causing materials shortages through a cradle-to-cradle approach to recycling. Communities are possible that use only renewable energy and recycle effectively. Wealthy societies could use taxation to curb excessive consumption and to generate the funding for a transition to a healthier society no longer powered by fossil fuels. Poorer countries may be able to skip some forms of wasteful growth and move directly toward sustainable systems with low environmental impact. But in every society, dialogue and compromises will be necessary. The transformations achieved will only win acceptance if they are explained, illustrated and made attractive and sensible through new narratives.

Should a society based on either no-growth or sustainability come to pass, however, some critics want more radical transformations. Many, such as Naomi Klein, see environmental issues through a Marxist lens and regard the current economic system as needing revolutionary change rather than reform around the edges. In *This Changes Everything* she argues that the climate crisis is not caused by humanity as a whole but by capitalism.[43] Klein focuses particularly on the fossil-fuel companies and their contributions to politicians and big environmental organizations, as well as disinformation activities and public relations campaigns designed to show how the market can best solve social problems, or why "green" projects are impractical and doomed to failure. She noted that mainstream news coverage of global warming had plummeted from 147 stories a year to just 14 between 2007 and 2011, which she explained as the result of a focused media strategy to push it into the background. As a Canadian, Klein gave particular attention to the protests of Alberta's indigenous population against the enormous project that extracts oil from tar sands. She reported stories from the Beaver Lake Cree Nation that some of the moose they hunted had inedible green flesh and cancerous tumors. The toxins in the air, earth, and water are not inevitable or unavoidable results of "progress." Rather, they arise from carelessness, cost cutting, greed, and ignorance, coupled with the capitalist drive to maximize profits. Klein sees hope in global resistance to the extraction agenda of corporations in the era of "tough oil," naming this movement the rise of "global blockadia."

Although Klein offers a heroic narrative of resistance to corporate villains, it is perhaps too easy to blame the environmental crisis on capitalism writ large. An oppositional critique simplifies the complicity of many world citizens. And how do people's lives unspool, day after day, after the global blockade ends? Arts and design also address practical problems associated with the current high carbon, high energy lifestyle of the global North. Journalist Kate Stohr and architect Cameron Sinclair titled their manifesto for humanitarian architecture *Design Like You Give a Damn*, rendering apparent the racism of indifference among designers and architects toward environmental suffering taking place in the global South. The nonprofit organization that Stohr and Sinclair led, Architecture for Humanity (1999–2015), catalyzed innovative designs and materials to house refugees and victims of "natural disasters." The environmental humanities might furnish a space where the global political critique of Klein and others on the ground is heard alongside stories of artists, designers, and makers of all kinds striving to make homes in this dangerous new climate.

There are massive challenges to stake an alternate future and no perfect model. Countries in the communist bloc had a worse record on pollution and misuse of energy than countries in the capitalist West. Moreover, the historical record suggests that it is inaccurate to see efforts to save the environment as a narrative in which government and industry refuse to reign in the exploitation of a resource until confronted by activists. In the thousand-year history of the North Atlantic fishing industry, for example, it was hardly so simple. In *The Mortal Sea* Jeffrey Bolster examines the history of fishing in the North Atlantic and shows that awareness of the dangers of overfishing were long understood. Yet on both sides of the Atlantic the same mistakes were continually made, despite persistent efforts to conserve marine resources. One species of fish after another was overfished and eliminated. Already in the seventeenth century, leaders in the American colonies imposed restrictions intended to protect fishing stocks, to little avail. Such government efforts continued throughout the nineteenth and twentieth centuries. Moreover, naturalists, journalists, and some commercial fishermen supported these efforts to prevent devastating overfishing, but poor practices continued. Bolster's conclusion might be applied to any number of other environmental problems in addition to fishing. He found that the European and American political and social systems were inadequate to the task, "with its checks and balances, its desire for prosperity and security, its willingness to honor a multiplicity of voices, its changing sense of 'normal,' and its shifting baselines, it was (and is) insufficiently nimble to stop the desecration of commonly held resources on which the long-term good of everyone depended (and depends)."[44]

In the not-too-distant future, people may well look at the present with wonder and disgust, much as people today look back at slavery. Why did entire societies refuse to develop or adopt indicators that included the environment as part of the economic health of a country? Why did so many leaders cling blindly to the ideology of endless growth, when anyone could see that the resources were finite? Why did human beings overfish the seas, devastate rain forests, and eliminate entire species? It would seem obvious that such behavior was perversely shortsighted and self-destructive.

There is another possibility, however, that our descendants will look back and see a successful deceleration of growth, coupled with a shift away from fossil fuels to alternative energies, away from extensive mining to cradle to cradle recycling, along with the expansion of regional economies based on localization. This may not be merely wishful thinking. In the United States, for example, the public has begun to demand "green homes." The *Wall Street Journal* reported that "green house projects" grew from just 2 percent of the market in 2005 to 20 percent in 2012, when it represented an investment of $25 billion. Projections are that this will rise to more than $100 billion by 2017.[45] Demand for "green" non-residential buildings is also increasing rapidly; in 2015 it accounted for 40 percent of the US market. "Green" buildings use 25 percent less energy, emit one third less greenhouse gas than conventional structures, and cost 19 percent less in maintenance.[46] Like the electrification of homes, which was a luxury for 5 percent of the American population for more than thirty years and then occurred rapidly between 1910 and 1940, it appears that solar panels, high quality insulation, and other "green" building materials may become widespread in the next two decades.

The United Nations adopted seventeen goals for sustainable development that went into effect in 2016. Goal 13 is to take urgent action on climate change, including a Green Capital Fund and the annual expenditure of $100 billion to mitigate CO_2 emissions, improve water quality, and focus on the environmental needs of developing countries, especially their rapidly growing cities.[47] These UN goals are not mandatory, but they were adopted unanimously by all 193 member states. One hopes they will be implemented more successfully than the 1992 climate accords signed by 116 countries in Rio de Janeiro, which contained many of the same goals, but failed to halt species extinction, slow global warming, or reduce poverty. But the sense of urgency has increased. The future might be one that embraces ecological citizenship formed according to what Ursula Heise calls an eco-cosmopolitan imaginary and understood as a creative materialist networking of human beings and all aspects of their environment. Such a future seems to demand a change in consciousness.

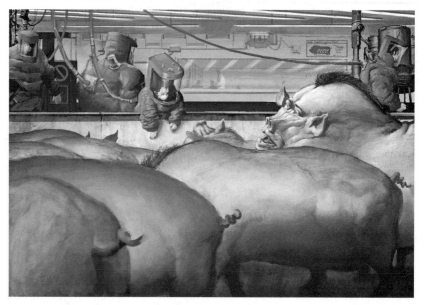

Figure 7.1
Jason Courtney's "Jimmy went in to see the pigoons," a painting inspired by Margaret Atwood's novel *Oryx and Crake* (McClelland & Stewart, 2003).

7 Unsettling the Human

New Theories

Suppose that all the movements and ideas discussed in the previous chapter were widely adopted. Suppose too that the degrowth movement made urban policy and that cities have become "green." Suppose that recycling has become biomimetic, reducing greatly the demands on resources. Will all then be well with the world? No, for these problems and solutions are focused primarily on the affluent West and on cities, and they do not take account of many environmental problems and much of the world. They are Eurocentric formulations, and some might say they represent a continuation of imperialism by other means. They are reforms, not a revolution. Moreover, as Adrian Parr has shown, the concept of sustainability has been hijacked during the past twenty years. Sustainability has become a corporate slogan and a government policy, and become commodified in the process. Companies cultivate a "green" image, the military seeks to achieve a "sustainable army," and celebrities drive electric cars, while the capitalist marketplace keeps expanding.[1]

This chapter turns to a range of challenging theories that have pushed the environmental humanities in new directions. As the strong presence of feminist political theory, anthropology, and international comparative studies should suggest, the environmental humanities are about more than reforming a wasteful economic system and adopting clean forms of energy. It is not only concerned to reduce carbon emissions, protect endangered species, and create better urban environments, worthy as all these goals are. These are adjustments and reforms that would permit the existing social and economic systems to persist without much change. But many scholars in the environmental humanities want to see more fundamental transformations: both in how people understand their place in the world as well as how social differences and interspecies hierarchies become "environmental" problems. Many theoretical developments are occurring simultaneously, and one chapter can only suggest the rich array of new ideas without trying to harmonize them into a single position. While there are many cross-cutting initiatives that might be included, we have elected to focus on new materialism, indigenous and postcolonial criticism, animal studies and queer ecology. These approaches all demand a radical redefinition of the relationship between human beings and nature.

New Materialism

One cluster of researchers links the field to continental postmodernist philosophy, calling for a new materialism that abolishes older distinctions between animate and inanimate matter. Inspired in many cases by Bruno Latour and by Michel Foucault, as well as by Karen Barad, Rosi Braidotti, Jane Bennett, and other present-day philosophers whose work is informed by quantum physics, microbiology, and the rhizomatic approach of Deleuze and Guattari, such researchers also claim inspiration from the technological transformation of matter through digitalization and the imbrication of human beings and computers in complex systems where the material world becomes an active part of cultural experience via digital media. Under the sign of the cyborg, one formulation of new materialism is indebted to Donna Haraway's critical feminist engagement with natural sciences.

The term "new materialism" arose during the late 1990s in philosophy, anthropology, and cultural studies, among scholars seeking a way through the impasse of modernism and postmodernism, displacing the centrality of the human subject in interpreting history, politics, and culture.[2] They have developed a somewhat insular technical vocabulary, including terms such as the posthuman, the more-than-human world, "becoming-with," vital matter, agentic realism, object-oriented-ontology or relational ontology, and onto-epistemic materialism. Yet new materialism is not just a new obscure fad, and this strange and sometimes rarefied philosophical discourse has something to offer non-specialists.

What features distinguish "new" from "old materialism?" And how are these ideas related to the environment? Of course, there is the colloquial use of "materialist," as someone who blindly pursues possessions, and there is philosophical materialism that rejects the action of supernatural or spiritual beings to investigate the natural world—human ideas included within it. But French intellectuals were rejecting neither of these traditions, but rather a so-called "vulgar" Marxist tradition, in which the material basis of society and class relations took such precedence over symbolic systems of exchange that human actions could seem almost mechanically predetermined. Some new materialists are primarily interested in how accounts of matter from quantum physics suggest agentic properties or an "active" role for what has been considered "inert matter." Objects, whether plastic cups, chairs, or objectified forms of other life, are seen as playing a more active role in human affairs. In comparison with twentieth-century Marxists, these thinkers place little emphasis on the rootedness of materials in class relations. According to the old materialist account, human labor, both manual and intellectual, transforms first nature into second nature, and the environment was where the work of extraction and transformation took place. Even for more nuanced Marxist critics and intellectual historians, such as Raymond Williams, the primary purpose of environmental cultural studies was to place excluded human subjects (rural workers) in the foreground to imagine new relations of society, culture, and environment.

New materialists are unlikely to view a cinderblock as the distillation of human labor, or analyze it in terms of exchange values. Instead, objects are framed first as actants—substances with some attributes of agency, or "agentic capacities." Objects are not conscious actors, but they are ineluctable parts of networks that human beings must master and use. Whereas a traditional Marxist critic might apprehend an object as a materialization of social relations, a new materialist sees the object as an actant within an assemblage. While relevant, social relations between human beings of different social status would not be privileged above relations between humans, other animals, plants, discrete technologies, weather, and so on. The new critical terms signal more than a semantic difference or a new jargon. Materiality is not conceived of as a mute, homogenous substrate nor as primarily an expression of the (more important) reproduction of social relations (as in Marx's "material production" or later, "material conditions"). For new materialists, the environment signifies not "nature" in the sense of a living system out in the world. Rather matter is an assemblage, a becoming-with-the-human, a "mesh" in which people are biologically entangled. Thus Timothy Morton emphasizes that ecology properly understood disturbs the old regime of mind over matter, nature against society.[3] Morton emphasizes throughout *The Ecological Thought* that we had better proceed with far more humility, a point seconded by historian Tim Lecain in his new materialist critique of the Anthropocene. "Rather than emphasizing human power and accomplishments," writes Lecain, "a neo-materialist view suggests that we are neither particularly powerful nor especially intelligent and creative—at least not on our own. Instead, the theory argues that we humans derive much of what we like to think of as *our* power, intelligence, and creativity, from the material things around us. Indeed, in many ways these things should be understood as constituting who we are."[4] For other contributors to new materialist thought, such as Kate Rigby, spirituality and cognition are also understood as immanent: arising with matter and through its surprising energies, and traceable through encounters with land, exemplified in Aboriginal concepts of country and care.[5] In their emphasis on immanence and the non-opposition of matter and spirit, many new materialists look to the ethics and metaphysical monism of Baruch Spinoza as a philosophical "road not taken," rejecting what they see as the destructive binaries of Cartesian metaphysics.

The environmental historian and political economist Jason W. Moore rejects Cartesian dualism in a wholesale critique of capitalism as a world-ecological regime. He distinguishes his materialist approach, or an ecological-dialectical view of nature-as-matrix, from a focus on metabolic rift in earlier Marxist environmental studies,[6] yet centers his thinking on a strong critique of capitalism. Moore rejects what he calls the "common sense of Green Arithmetic," according to which "Society plus Nature equals environmental studies." Instead, he urges us to conceive of "a nature that operates not only outside and inside our bodies (from global climate to the micro-biome) but also *through* our bodies, including our embodied minds."[7] Moore applies this ecological-dialectical materialism to analyze capitalism in the twenty-first century, arguing that the era of Cheap Nature is over and that global warming likely precludes a second "Green" capitalist agricultural revolution.[8] "The problem today," he writes, "is one of capitalism exhausting its *longue durée* ecological regime."[9] *Capitalism in the Web of Life* brings into focus a material horizon to capitalism as a way to organize nature and a dominant way of valuing the web of life.

In *Vibrant Matter*, Jane Bennett seeks to "to dissipate the onto-theological binaries of life/matter, human/animal, will/determination, and organic/

inorganic" and instead develops a "notion of publics as human-nonhuman collectives that are provoked into existence by a shared experience of harm," such as the experience of the massive electrical blackout in 2003.[10] This approach has affinities with constructivist social science, insofar as Bennett moves away from a politics of blame, in which individuals alone are responsible for events. Preferring a view of agency as distributed and arising from swarms, networks, and assemblages, she distinguishes between "vital materialism" and phenomenology and social science that emphasizes the power of structures and contexts.[11] Considering food as an actant, she writes, "call the assemblage formed by these human and nonhuman bodies 'American consumption' and name as one of its effects the 'crisis of obesity.'"[12] Likewise, one might see high energy transportation as an effect of the assemblage or infrastructure of human and non-human elements that make up a traffic system. In short, Bennett advocates a theory of democracy in which "human culture is inextricably enmeshed with vibrant, nonhuman agencies" and therefore calls for "new procedures, technologies, and regimes of perception that enable us to consult nonhumans more closely."[13] Building on Karen Barad's concept of a public as an interaction of humans and non-humans, Bennett's materialist politics does not privilege people or other life forms over the inanimate.

This approach is kin to actor-network theory, and its value lies in developing ways to see people not as sovereign individuals but as parts of complex systems that are structured to move in certain directions and to express certain values. However, its focus on erasing the line between the human and materiality opens a procedural can of worms, or the problem of how to include the agency of animals, fish, birds, or other non-human life forms as well as natural forces and weather cycles. Bruno Latour confronts this directly in *Politics of Nature*, where he calls for a political ecology that assembles a new body politic of human and nonhuman actors of the "pluriverse"—though that new Constitution hardly ends metaphysical disputes over human, non-human, and nature.[14] Bennett also pitches agency in this positive, constructive sense, as being enacted through a human/ non-human assemblage. Her work is not about people being dominated by systems or merging into an unconscious reactivity in networks that hold them captive, like the society depicted in the *Matrix* films. Rather, Bennett seeks to arouse the reader's "sensibility," or change the sense of one's relation to the world, and she hopes that others will tease out the fuller implications of her vital materialism for "ecohealth."[15] Readers need to feel that they are "Earthlings as well as women, men, Americans, Republicans, et cetera. Nudging people to feel more of the rich complexity of material life, to think of nature not as something out there but as a set of interacting forces, flows, and entities at work inside our bodies as they also form various kinds of links across bodies." As she told an interviewer: "If you want to move so-called representative democracies toward more sustainable modes of consumption and production, the demos has got to want it."[16]

Conceiving of politics as decentralized and de-centering movement, without a claim to sovereignty or a single core identity, seems to be one promise of a new materialism. Ecologizing politics in this way means seeing humans not as individual citizen-subjects who decide by reasoning or limiting their self-interests in some hypothetical primal scene of social contract as conceived by liberal theorists. Rather, politics becomes a less predictable activity of porous bodies subject to contagions of mind and feeling as well as physical vulnerability. This is the analytical line of thought pursued by Heather Houser, for example, in her study of ecosickness narratives in recent film, fiction, and nonfiction. Sickness linked to uncertain environmental causes—whether heavy metal pollution, radiation, or exposure to endocrine-disrupting synthetic chemicals—Houser takes as a road to insight into "the imbrication of human and environment."[17] Specifically, by reading the complex emotional responses and narrative forms that have preceded regulatory action on environmental toxins, Houser describes how "the embodied person is enmeshed in macro processes of technologization and environmental manipulation," an "enmeshment [that] does not dictate a singular ethics or politics."[18] Reading networks of perception and feeling alongside enviro-technological assemblies of health, such as animal testing laboratories and clinics who treat patients with multiple chemical sensitivity, thus becomes a step toward a politics of human and ecological health.

A subset of new materialism goes by the rather infelicitous label "material ecocriticism."[19] Ecocriticism, or environmentally oriented and self-reflexive literary criticism, has benefited by a refreshing turn to materiality in its analysis of cultural phenomena. What is "new" in the new materialist ecocritical turn is the sophistication of critics' concept of matter as continuous with structures of feeling and power. With this has come an opening of critical attention to wider aesthetic categories (including filth, waste, noise, ugliness) inherent in a less idealized, pure sense of environmentality.[20]

Moreover, a framework of material continuity or philosophical monism recognizes literary and other cultural texts as resulting from processes continuous with material formation of human lives and the creation of landscapes and urban forms. They share this with some human geographers, travel writers, and ethnographers who are crossing the species divide in their work.

The practice of material ecocritical reading shuttles between literary texts, landscapes, and encounters in specific places; a forest can be read alongside a poem. Likewise, narrative nonfiction can explore the "deep map" of a landscape, as Loren Eiseley did in *The Immense Journey*, which combined an intimate acquaintance with the geology, evolutionary biology and history of the Great Plains. Susan Naramore Maher examines the "literary cartography" of such writers, who move beyond the immediately visible monoculture of modern agriculture to recover the biome of the tall grass prairie, the shaping force of the glaciers, the ancient seas, and the forms of life that disappeared with the fifth great extinction.[21] In such writing, the land is not ultimately the object of our control, but a changing ecological system of living matter and signs. Interpreting its dynamism requires ecocritics, multispecies ethnographers, biosemioticians, environmental historians and perhaps ecopsychologists to understand each particular landscape's agency in human history.

First Peoples

Another group of scholars in postcolonial studies, ecocriticism, and anthropology focuses on how First Peoples inhabit the world. Working with indigenous co-producers of knowledge, these researchers seek to understand non-Western sensibilities that have never divided culture from nature. Indeed, some humanists were first drawn to consider environmental questions after extended time living among First Peoples. Whether studying their language, rituals, food ways, or social structure, this tradition has certain affinities to new materialism. Animist ethics and the idea of lively matter are kindred ideas. Each rejects market capitalism and modernity as inadequate, and in different ways each advocates a different structure of consciousness. New materialism also puts human bodies alongside other living organisms in a more horizontal fashion, a point of convergence between posthumanists, indigenous and animals studies scholars and a central intervention of feminist critique. Finally, the new materialists try to construct an alternative phenomenological description of the world, while postcolonial and anthropological approaches seek to understand the life-worlds of peoples who retain a non-Western sensibility.

An essential aspect of non-industrial cultures often is a conception of language not as a grammatical system but as an expression of the relation-ship between human beings and place. Environmental philosophers have moved to recover this connection. As the philosopher Jim Cheney put it, "Though the epistemologies of modernism detached themselves from the world—treating the nonhuman world and even the human world as objects of domination and control—and though the postmodern view of language and self (the self as solipsistic maker of worlds) to a large extent reflects this detachment, we and our languages are fundamentally of the world."[22] He argues that language is locatable, connected to place. Narra-tive performs the necessary function of connecting us to a specific world, and in his view the idea of creating a value neutral language is a Western illusion. Native American "knowledge of the natural world, as I have said, is based on an epistemology of respect requiring attentive listening to, and reciprocal communication with, the earth and is woven together within ceremonial worlds designed to accommodate human culture within (and as) a wild world. This knowledge is, essentially, a comedic way of being in the natural world rather than a tragic separation and alienation from that world."[23] Overcoming that sense of alienation, embodied in the value neutral language of modernist science, is an essential act of recovery that underlies the environmental humanities. Likewise, indigenous studies are enriched through forms of ecological knowledge and value that may come only through appreciating multilingual cognition and preserving endan-gered native languages.

Depth and archetypal psychology as well as postcolonial criticism offer more than functional or biochemical accounts of how individual psyche, social script, and the natural world come together. Ecopsychologists seem more willing than others to conceive of nature as a world of living enti-ties and symbols that preceded and will survive human society. The eco-critic Rinda West, for example, deploys Carl Jung's concept of the shadow to argue forcefully against dominant structures of Western thought that repress alternative relationships of humans and nature. "The stories we hear most often," writes West, "alienate all of us from nature and from our own sympathetic responses."[24] Human experiences of alienation, vulnerability in nature, and connection to land nonetheless resurface from the shadow world of the unconscious. But without an interpretive process or alternative guiding stories, many individuals in the West experience the real domination of the natural world, other animals, and native peoples as anxiety and a "quarantining of psychic energies."[25] West turns to recent novels, particularly those written by Native American authors including Leslie Marmon Silko, Louise Erdrich, and N. Scott Momaday, but also work by Margaret Atwood, Marilynne Robinson, and feminist ethicists who offer "alternative narratives that imagine a different relationship between humans and the land," and "stories about connection, community, and personal well-being."[26] Perhaps such stories are more needed because they contradict the present-day drift of environmental change.

What does it feel like to contemplate a proposal to build an eighteen-story telescope observatory on your people's sacred summit, a mountain so divine it is revered as the Sky-Father and origin of life (Mauna Kea, Hawaii)? Or to picture burial grounds buried in mine slurry (Beaver Lake Cree, Alberta, Canada), villages contaminated by oil spills and improperly capped wells (the Cofán, Siona, and Huaorani of the Oriente, in Ecuador and Peru; the Ogoni people of the Niger River Delta)? Or to lose children to poisoning by arsenic byproducts of gold mining knowingly released into your community's streams (the Dene of Yellowknife, Canada)? For many indigenous groups, or 5 percent of the world population, sacred lands have an absolute value that cannot be exchanged or repaid. Thus, development by corporations and governments (often directly descended from colonial forces) represents more than an assault on native peoples' resource base. The sociologist Al Gedicks has argued that the closeness of native peoples to their land—physical and metaphysical—makes them not only "vulnerable to change in their ecosystems" at a subsistence level. Ecological damage also threatens their cultural survival.[27]

Indigenous and postcolonial studies begin with a critique of colonial history as social-ecological violence perpetrated against lands and peoples. The anti-imperialist critic Walter Rodney put the first step unequivocally into the title of his 1972 classic *How Europe Underdeveloped Africa.*[28] As industrialized Western empires expanded, they appropriated the biological and mineral wealth of dispossessed "ecosystem people" and incorporated surviving indigenous groups as workers and consumers.[29] What was

unthinkable within a native environmental ethic and cosmology was perpetrated through violent conquest and appropriation of tribal lands by
forced treaties. By the end of the last century, the abstract idea of land
as property, to be strip-mined and clear-cut, came to dominate much of
the face of the earth. Even where Western conservationists created colonial
national parks, these came with a systematic devaluation of local knowledge and often failed to protect threatened species. Reversing the colonial
value system is often a precursor to proposing alternatives to a top-down
land management model. Clapperton Mavhunga describes in a case study
of African indigenous responses to Zimbabwean conservation the "technologies of everyday innovation," including use of cyanide-laced fruit to
hunt game, insisting that technologies to shape livelihoods from forests
and farms are not something brought from outside.[30] Mavhunga predicts
that until governments and development agencies conceive of ordinary
people as having creative agency in their everyday lives, efforts to conserve
species or "develop" indigenous lands will fail.

Legacies of environmental degradation via colonization cannot be
unthought by present-day native peoples. They inform what the Spanish
economist Joan Martinez-Alier calls the "environmentalism of the poor,"
who see action on behalf of environmental quality not as a luxury but
as a fundamental right of all people.[31] Such environmental thought and
political movements are rooted in common vulnerability and represent
demands by indigenous and poor people, who in many countries represent
a majority.[32] This situation presents an obligation and conceptual opening
for the emerging environmental humanities, not least because it helps to
internationalize a field that otherwise could become Eurocentric or focused
on North America. Comparative studies such as Nixon's *Slow Violence* have
launched conversations about how to further democratize teaching and
scholarship to represent all peoples.

In late twentieth-century North America, indigenous-led environmental
movements have asserted tribal rights to land framed in terms of environmental justice. Indigenous groups often make a dual claim to environmental justice on the basis of procedural and distributional claims.[33] First, lands
and waters were taken through often illegal procedures, either through violent expulsion of resident peoples or dishonored treaties. Second, individual groups have appealed for a re-distribution of environmental amenities
and risks, after suffering disproportionate exposure to toxins, for example

by living downwind and downstream of mines[34] or near open air atomic tests.[35] An alliance of Native groups in California, the InterTribal Sinkyone Wilderness Council, successfully litigated to protect traditional tribal lands. In the case of the Sally Bell Grove, a forest eventually became a partnership between a private land trust and the Council for wilderness conservation, cultural preservation, and education within a tribe's historical range.[36] What began as a regional alliance and one successful conservation trust has become a model for national cooperation between Native groups and the Trust for Public Land. Their partnership represents both ecological restoration and cultural healing, where autonomy *and* environmental protection are recognized.

One outcome of the alliance politics of indigenous political ecology is a widening and deepening of the environmental justice narrative. Environmental justice historically has been a conceptual weapon of the weak, first wielded by African-American and Latino/a organizers against chemical waste, agricultural toxins, and "dumping" on poor communities. And as Joni Adamson, Rachel Stein and others have shown, everyday environmental justice activism has deep roots in indigenous communities.[37] Naomi Klein has remarked on the leading role played by indigenous groups in growing environmentalist movements that have united exploited resource workers, feminists, and farmers to block further extraction of fossil fuels.[38] This coalition does not assert a singular identity based in land, blood, or language, but acts together to protect a despoiled planet and human health. Such a coalition is also at work in the opposition to the Keystone XL Pipeline from Alberta, Canada through Montana, South Dakota, and Nebraska, running through Native American lands (notably the Lakota Sioux), including sources of water, sacred places, and graves.

Resistance to oil pipelines exemplifies the struggle for control in decision making, relinking land, people, language, and heritage, to reaffirm fundamental relations of life and livelihood. Negotiating sovereignty for native tribes on their land means shifting authority to manage resources but it does not automatically result in ecological health or sustainability. Native groups may decide to allow mining or drilling for oil and gas, to accept nuclear waste facilities, or to continue hunting of prized species like gray whales, beluga, or black rhino. A moralizing, Manichean concept of ecological Indians would label such choices "not authentically native," however damaging they may be. Some native leaders have ignored recommendations of environmentalists and conservation biologists and developed resources on their land, at times to alleviate community poverty but sometimes for personal gain.

Many conservationists and environmentalists have been outraged by native groups' decisions to hunt species threatened by extinction. Indigenous scholar Charlotte Coté defends the Makah tribe's decision to hunt gray whale when they were removed from the Endangered Species List in 1994, explaining that whaling "reaffirm[s] our identities as whaling people, enriching and strengthening our communities by reinforcing a sense of cultural pride."[39] Among the Inuit and other indigenous peoples of the circumpolar Arctic, hunting traditional foods (including whales) is central to cultural identity.[40] Cultural, spiritual, and tribal identity have all been threatened by the market-oriented industrial fisheries of the West, creating a situation where indigenous groups who hold prey animals sacred and feel a duty to hunt them have found themselves at odds with conservation biologists and environmentalists. The convergence of historical injustices, climate-driven resource pressures, and the need for deep cultural and linguistic translation places indigenous studies at the cutting edge of central issues in the environmental humanities.

Humanists must articulate ecological values in relation to indigeneity with historical alertness and political sensitivity. As was discussed in chapter 2 in conjunction with the emergence of "new wilds," archeological research has revealed that Native environmental modifications were undertaken at a much vaster scale than previously thought. From William Bartram's description of the people and landscapes of the Southeast, we have written evidence of extensive cultivated fields, raised mounds and town sites.[41] These depended on organized, coordinated management of forests, either by girdling undesirable trees (favoring sweet-acorn white oaks at the expense of red oaks) or allowing desirable fruit trees to spread through seed dispersal into crop areas. Native peoples shaped local and perhaps regional vegetation cover in the Americas.[42] Moreover, the environmental philosopher J. Baird Callicott has explored how agro-ecological practices from Amazonia to the Eastern Woodland peoples of North America were nested within worldviews and knowledge of systematic connections that are rightly characterized as "ecological."[43] Callicott points out that Polynesian genealogical chants even present creation proceeding upward in evolutionary terms, from coral polyps to marine worms and so on.[44] Much can be learned from comparative, synthetic studies such as Callicott's, for they raise historical questions as well as philosophical and ethical ones. Notwithstanding their beliefs, were native peoples on the whole ecologically benign in their effects?

In *The Ecological Indian: Myth and History*, the anthropologist Shepard Krech, III challenged the stereotype that Native peoples were ecologically minded or conservationists. He does recognize the strong alliance of present-day environmentalists and indigenous groups and the deep ecological knowledge and value of interspecies relatedness across many groups. But *The Ecological Indian* also debunked an idealized, ahistorical view that concluded that *because* indigenous groups held proto-ecological beliefs, they also restrained hunting and gathering or that they practiced farming and set fires as modern conservationists would recommend.[45] Drawing on a vast but fragmented archive, Krech found that not all historical environmental changes attributable to indigenous groups were benign by today's standards of conservation biology. The scathing reception of *The Ecological Indian* by scholars and activists in Indigenous studies reveals a widespread drive to preserve the myth of noble, ecological savages—a myth that is ultimately dehumanizing. Yet many accused Krech of being "anti-Indian."[46] Kimberly TallBear opened an alternate path in her response to Krech's work: "The fundamental difference between natives and Euro-Americans lies not in being ecological saints versus being the ecological anti-Christ. In my mind, the principle difference between us lies in the difference between grappling with and throwing off the yoke of colonization versus coming from a culture that was born of and grew from the fruits of colonization."[47] The environmental humanities must support the difficult work of grappling with forms of colonial oppression and the history of indigenous groups, while recognizing that history itself is a Western construct and way of knowing that must take care not to ignore other forms of knowledge.

Cosmology and ideology do not automatically determine actions and behaviors. How then to conduct the work of translation and the insider-outsider transaction that is necessary for indigenous studies to have the greatest impact as conduits of worldviews and values? One must combine a political openness of affiliation (not to be confused with neutrality or disengagement) and participatory research methods (not "ethnography light") that invest real time and energy when crossing social boundaries. The environmental humanities might champion not "becoming native" as a process of acquiring authenticity or stepping out of history, but rather as a process of intercultural learning, as an extension of affiliation, kinship, care, protection, wonder, curiosity, generosity, and reciprocity. Ethnography represents the deepest investment of a researcher's time, training, and patience: it is unsettling of disciplines and self. The result of training is the ethnographer's consciousness itself, a self-reflexive tool honed in the co-production of knowledge with land-based partners. Meaning does not come cheap. When researchers are just passing through or aiming to exploit sacred or scientific knowledge for academic or commercial gain, local councils and authorities will resist cooperating with them. The record of Western exploitation justifies this wariness.

Animal Rights

Another outcome of the struggle to encounter indigenous thinking and imagination is the posthuman and ontological turn in anthropology and cultural studies. Here, the "human being" comes apart along with the hermetic notion of academic reasoning into many contested versions of the human that creep, swim, or strut abroad. The final sections of this chapter explore two welcome rethinkings of the "human" in "environmental humanities": research considering human-animals among and with other species; and queer ecology, exploring the gendered, performative construction of environmental politics, ethics, the heteronormative reproductive mode of environmental futurity, queer empathy with ecological others, and more. It is no accident that opening the complexity of human sexuality came with the first ethnographic encounters of Margaret Mead, Gregory Bateson, and others living among First Peoples. Queer theorists and animal studies scholars are opening up mental categories, environmental ethics, and politics as projects of selves and societies by giving priority to relationality. The philosopher Emmanuel Lévinas also privileged relation and therefore placed ethics as a primary condition for ontology.[48] Before we become who we are, we are already born in relation to a series of others. In particular, Lévinas posited that we become ourselves by turning toward the other. The "other" of Western moderns is the person deprived of resources and political recognition: the illegal migrant worker, the refugee, the ethnic or religious minority. For Lévinas, the other is always human, a person we can meet face to face, but encounters with other animals may provoke a more radical estrangement of self-interests—and therefore a more radical environmental ethics and politics. Consider the poet Walt Whitman's lines on contemplating domestic animals in "Song of Myself": "Not one is dissatisfied—not one is demented with the mania of owning things; / Not one kneels to another, nor to his kind that lived thousands of years ago; / Not one is respectable or industrious over the whole earth. / So they show their relations to me and I accept them, / They bring me tokens of myself, they evince them plainly in their possession."[49] Whitman ends the stanza by wondering about this relationality between himself and other animals and gropes toward an evolutionary connection: "I wonder where they get those tokens," he asks, "Did I pass that way huge times ago and negligently drop them?" The question frames becoming human as a path of loss and forgetting rather than progression upward into consciousness. An avid reader of Darwin, he discovers in contemplating animals not himself as "Man the Master" at the head of a great chain of being, but his animal similitude as a negation of the dominant mode of being human. Whitman diagnosed nineteenth-century society as being ill with sickening dissatisfaction, manic possessiveness, and obsessive moralizing. More specifically, he turns to animal-being as an alternative to middle-class respectability and the Protestant work ethic—and Whitman is certainly not the first or last poet to apprehend in animals sentience, emotions, and even a kind of instinctive wisdom.

At the moment when some scientists want to treat animals as raw DNA material to be manipulated, other scientists, many in the environmental humanities, and most First Peoples assert that animals are intelligent and emotional beings with as much right to exist as humans. Some activists feel strongly about this and demonstrate outside scientific laboratories, which they regard as prisons where animals are mutilated and tortured. Likewise, keeping animals captive in traditional zoos has in many places been replaced by more spacious facilities that attempt to provide appropriate environments, informed by ethologists, zoologists, and more recently, biosemioticians. Such long-term changes in how humans treat other animals have been interpreted as a progressive ethical process, where we become more humane—even human—through our engagements with animal others.

Val Plumwood, one of the founders of the environmental humanities, was a staunch critic of Western anthropocentrism. She developed the concept of "hyperseparation" to describe the binary ideology that divided human beings from nature and defined it as subservient and secondary. She was not the first to attack Cartesian dualism, but she was one of the first to show how a wide range of "others" were systematically denigrated by identifying them with nature, including women, the colonized, the indigenous, and the non-human. The domination of nature was linked to the subjugation of many groups. Plumwood's later work focused increasingly on the dualism of mind and matter, which granted special privilege to (the implicitly male) mind and depicted the rest of the world as mindless material. This view stimulated her interest in the philosophical ecology of Aboriginal Australians, who viewed all animal life as sentient. Secular philosophy and early anthropology rejected such thinking and labeled it "animism." It was regarded as primitive and superstitious, in comparison with the cool objectivity of science. Plumwood saw animals much as Native Americans or Aborigines do, as "kindred beings." The Sioux holy man Black Elk likewise spoke of "the two legged and the four legged" as relatives, and the Ojibwa see a wide range of "persons," only one of which is humankind. Only in quite recent times have scientists begun to understand that chimpanzees can learn American sign language or that the songs of whales and birds might be meaningful communication. Once this is admitted, it follows that groups of animals can also have cultures of their own, whether a herd of elephants, a pod of whales, or a flock of birds. It has been documented that two species of dolphin have complex and stable cultures that include using tools, abstract concept formation, self-awareness, and vocal communication.[50] Their behavior cannot be explained genetically but is best explained as learned, cultural behavior. As one Aboriginal Elder put it, "birds got ceremony of their own—brolga, turkey, crow, hawk, white and black cockatoo—all got ceremony, women's side, men's side, … everything."[51]

Science itself has increasingly recognized that animals can be said to have emotional lives, though ethicists continue to debate which animals have consciousness, to what degree, and which criteria are indeed pertinent to recognizing the inherent value of other creatures. The philosopher Tom Regan, for example, proposed that animals in possession of complex awareness, as "subjects-of-a-life," must be accorded inherent value—the *a priori* right to not be used as mere meat by humans.[52] Warwick Fox has responded by denying that other animals share the human "mindscape," though he would extend human protection to endangered species and humane criteria of treatment to domestic and wild animals—as a principal of a responsive, ethical culture.[53] Notwithstanding the controversy over the extent of animal consciousness, it is indisputable that our animal kin experience things we cannot. Many animals have more acute senses than people do. Dogs can hear pitches much higher than human beings, and they can be trained to smell illegal drugs in an airport or the beginnings of cancer in a patient. Cats have night vision far superior to human beings, and birds navigate thousands of miles in annual migrations. A leading philosopher of Animal Rights, Peter Singer, Professor of Bioethics at Princeton University, argues that the vital question to ask is not whether animals can talk or reason but whether they can suffer. Descartes famously argued that animals did not suffer, but few would agree today. Animals have nervous systems quite similar to those of human beings. Singer develops Jeremy Bentham's arguments for treating slaves and children more humanely and applies them to the treatment of animals. He declared, "If possessing a higher degree of intelligence does not entitle one human to use another for his or her own ends, how can it entitle humans to exploit non-humans?"

Singer also develops an array of practical arguments against eating meat. He points out that there are three times as many domestic animals as human beings, raised and fed using industrial methods that consume large stocks of fossil fuel and chemical fertilizers. Moreover, "Since 1960, 25 percent of the forests of Central America have been cleared for cattle. Once cleared, the poor soils will support grazing for a few years; then the graziers must move on. Shrub takes over the abandoned pasture, but the forest does not return. When the forests are cleared so the cattle can graze, billions of tons of carbon dioxide are released into the atmosphere. Finally, the world's cattle are thought to produce about 20 percent of the methane released into the atmosphere, and methane traps twenty-five times as much heat from the sun as carbon dioxide."[54] In short, not only do animals have rights, but their mistreatment is a central part of the environmental crisis. Singer published *Animal Rights* in 1975. At the time, DNA had been discovered, but synthesizing life was only science fiction. As might be expected, neither Singer nor other advocates of animal rights support the creation of synthetic species, even if the goal is not financial gain. Some abilities have now been transferred from one species to another. Modified cows supply plasma for human blood transfusions. Salmon do not freeze at low water temperatures, and the relevant salmon gene has been planted in selected fruits and vegetables.[55] Just as coal can be made into plastic, animals and plants are being treated as raw materials to be transformed into beings never found in nature, in ways that animal rights activists regard as grotesque and violent. Bioengineering is erasing the line between animal and human, so that the future of medical treatments likely involves phenomena such as pigs raised with human kidneys ready for donation—further eroding the boundaries of the human. Experimentation on animals figures heavily in Margaret Atwood's dystopian novel *Oryx and Crake*. Atwood imagines a near future in which commercial farms raise transgenic pigs for organ transplants. The brand-named "pigoons" have proprietary genetic code, spliced with human and baboon DNA. Inspired by the novel, the artist Jason Courtney has painted an unsettling encounter: inside a futuristic lab/pen, a man-faced pigoon rears up to face a boy engulfed in an oversized Hazmat suit. (See figure 7.1.) Concentrated animal feedlots already resemble the pigoon factory; they are breeding grounds for cross species transmission and anti-biotic-resistant pathogens. But there are increasing numbers of people who push back against this instrumental use of animals, becoming vegetarians and refusing to use leather goods or to buy products tested on animals.[56] As Lisa Kemmerer has argued, the typical flesh eater consumes 2,600 creatures in a lifetime, most of them female. "They are genetically manipulated, warehoused, and transported as if they were objects—stock—rather than sentient individuals."[57]

Queer Ecology

The conceptual linkages are even more braided between material ecocriticism and the critique of bodies, power, and sexuality in feminist and queer theory. Some of the strongest research on culture, literature, and environment, such as Stacy Alaimo's *Bodily Natures*, re-interprets the meaning and power of human bodies in relation to social categories and environmental endangerment and develops from sustained attention to the body in feminist theory.[58] Alaimo's concept of *transcorporeality* captures in one word a different paradigm for viewing the human: as porous bodies infiltrated by toxins and cultures for microbes, we are inherently part, product, and actor on behalf of our environment.

The transcorporal human suggests an alternate focus for a new kind of science, medicine, and humanities—a feminist, perhaps queer subject of environmental knowledge. Feminism is not a single ideology, but a discourse, and as such hard to summarize. But there are some shared points of view. As Evelyn Fox Keller emphasized, traditional science, as developed primarily by white males, treats nature as an alien other to be observed, manipulated, and controlled. As an alternative, Fox Keller posited "dynamic objectivity":

Dynamic objectivity aims at a form of knowledge that grants to the world around us its independent integrity but does so in a way that remains cognizant of, indeed relies on, our connectivity with that world. In this, dynamic objectivity is not unlike empathy, a form of knowledge of other persons that draws explicitly on the commonality of feelings and experience in order to enrich one's understanding of another in his or her own right. ... Dynamic objectivity is thus a pursuit of knowledge that makes use of subjective experience (Piaget calls it consciousness of self) in the interests of a more effective objectivity.[59]

Timothy Morton has observed that at first glance it would seem that queer theory and environmentalism make impossible bedfellows. He argues that "Ecofeminism (the classic example is Carolyn Merchant's *The Death of Nature*) arose out of feminist separatism, wedded to a biological essentialism that, strategic or not, is grounded on binary difference and thus unhelpful for the kinds of difference multiplication that is queer theory's brilliance." He continues: "Much American ecocriticism is a vector for various masculinity memes, including rugged individualism, a phallic authoritarian sublime, and an allergy to femininity in all its forms."[60] Morton is certainly not the first to note the gendered dynamic of wilderness-oriented ecocriticism in North America; but he overlooks the conceptual influence of ecofeminism on queer ecology. Such formative early work as Greta Gaard's "Toward a Queer Ecofeminism" is notable in its *non*-essentialist, coalitional politics, seeking common ground for queer and ecofeminist theory precisely in a non-dualistic approach that nonetheless opposes masculine domination.[61] Morton's high-profile 2010 polemic acknowledged Gaard and Catriona Sandilands as having done "pathbreaking work," but neglected to note that their innovations anticipated some of his arguments.

Queer ecology, or interpretive research on the intersection of environmental issues and the full range of human sexuality, emerged almost simultaneously across geography, cultural studies, ecocriticism, and anthropology. The ecocritic Catriona Sandilands describes queer ecology as sharing an agenda with environmental justice activism as "marginalized communities crafting new cultures of nature against dominant social and ecological relations of late capitalism."[62] These relations include sexuality. Forms of desire, sexual partnering, and politics of identity are naturalized, commodified, and granted (or denied) environments of expression. Our deepest phobias and categories of good, beautiful, ugly, and bad environments are often coded with sexual identity and connotations.[63] Other humanists have observed that sexuality in nature is polymorphous, with plants and animals exhibiting asexual reproduction, bisexuality, and homosexual and heterosexual pairing. Viewing heterosexual monogamy as "natural" ignores the complexities of biology: "There's no contradiction between straightforward biology and queer theory," Morton puts it. "If you want a queer monument, look around you."[64] Yet it would be reductive to argue that biology can explain human sexuality, for the play of erotic desire is also a matter of culture and aesthetics, of place and timing.

A signature move in queer ecology is to decouple pleasure and desire from the potentially destructive logic of heteronormative reproductivity. One strand of thought has built on Lee Edelman's critique of "reproductive futurism," the monopoly on future-oriented ethical and political claims wielded by heteronormative societies. These are organized around sentimental representations of "the child" and hold childbearing as the prime social duty.[65] The literary critic Robert Azzarello has argued, for example, that a narrowly reproductive sexuality may contribute to environmental problems—not only through human population growth but also through a belief that all environmental harms can be fixed through a self-reproducing world. In the place of a heteronormative, reproductive stance toward nature, Azzarello endorses what he calls "queer environmentality," or "a habit of thought that conceptualizes human beings, other life forms, and their environments as disregarding ... the ostensibly primary, natural law 'to survive and reproduce.'"[66] In work that is primarily reparative and revisionary, he interprets the non-conformist sensuality of writers (among them Thoreau and Melville) whose lives and works did not conform to the family-oriented morality of their day.

The ecocritic Nicole Seymour has pushed back on this anti-futurity strain in queer ecology in *Strange Natures: Futurity, Empathy, and the Queer Ecological Imagination*. She explores how recent queer novels and films have represented ethical relations between queers and non-humans in part to describe a different paradigm for caring about the environment, the future, and nonhuman life, as well as concerns of environmental justice that are configured by sexuality as well as gender, race, and class. Why should caring be assigned to a narrow conception of sexuality and gender roles, in particular to mothers as archetypal protectors against environmental threats?[67] What does it mean to articulate "queer ecological concerns?" And why might this be important? Seymour proposes that recent novels such as Leslie Feinberg's *Stone Butch Blues* and Shelley Jackson's *Half Life* and films such as Todd Haynes's *Safe* and Ang Lee's *Brokeback Mountain* ultimately teach queer values that are biocentric: "caring not (just) about the individual, the family, or one's descendants, but about the Other species and persons to whom one has no immediate relations."[68] Queer ecological reading intervenes not only at the ethical level but also through the aesthetics of camp. *Strange Natures* is optimistic about how queer aesthetics might revitalize environmentalism by shaking off its puritanical earnestness, inserting playful humor and self-deprecating irony. If one of the main concerns of environmental humanities is how to care for a post-natural, damaged planet riven by social inequities that drive further damage, this mode of empathetic reading will be central.

Queer ecology also nudges us to notice the plural desires projected onto wild areas. For example, the same city park can be a site of night time cruising, daytime picnics, lunch breaks, and ongoing ecological research. The geographer Matthew Gandy explores the case of Abney Park in North London, an overgrown nineteenth-century cemetery that has become a popular destination for a wide range of visitors: "artists, cruisers, dog walkers, drinkers, ecologists, joggers, lovers, mourners, photographers, poets, writers, and many others."[69] Gandy brings together the insights of urban ecology—particularly of urban political ecology which recognizes value in unplanned wild areas in derelict industrial sites—and queer and posthuman theories of space. The latter treat desire as crossing "beyond individual or even multiple human bodies to incorporate nonhuman nature, inanimate objects, surfaces, and smells."[70] Queering urban nature is not only about recognizing the ubiquity of non-heteronormative sexuality and erotic behaviors in the actual use of public spaces. It is also a matter of a conceptual synergy between queer spaces and the alliances that emerge to protect specific sites—in the case of Abney Park, a place recognized officially for its biodiversity is also internationally known as a destination for cruising gay men. The park's plural meanings depend on it remaining an open safe haven.

Queer ecology, animal studies, indigenous studies, and new material-ism enrich our understanding of the plurality of human experiences of the surprising, vital world. They demand we re-think the grounds of our environment-altering humanity. They challenge narrow and hidebound ways of construing the purpose of culture, society, and politics relative to environmental problems. And they challenge pseudo-evolutionary argu-ments about the naturalness of social orders or divisions between human and nonhuman animals. David Abram in the "Commonwealth of Breath," describes the primacy of breath common in oral traditions, evoking in impressionistic terms how "for our oral elders and ancestors, that which dissipates as smoke or dissolves into the unseen air is by that very process slipping *into* the mind, binding itself back into the encompassing aware-ness from which our bodies steadily drink, the wild sentience of the world, moody with weather..."[71] According to Abram, waking to our participation in the commonwealth of breath is step one to acting on climate change. Through the new theories and approaches presented in this chapter, the environmental humanities are enlivening our apprehension of the hetero-geneity and wildness of sentience. Through their engagements with the arts and lively everyday creativity, they bring us to our senses.

Figure 8.1
A photograph of Ohio or "Sunshine" Key taken by Flip Schulke in June 1973. Rachel Carson lived there while gathering data for her 1955 book *The Edge of the Sea*. Since then, it has been dredged, filled, and paved over to make a trailer and RV park. Source: Documerica Series, Record Group 412, Arc – 548632, National Archives

8 Conclusions

Restructuring Knowledge

The environmental humanities are a new formation in the ongoing development of universities, and they may also represent a deeper shift in the organization of knowledge practices. The global problems discussed throughout this volume challenge scholars to redefine inherited disciplines, often by working in larger teams with artists, practitioners, and entrepreneurs. Ethics has become indispensable to medicine and biology. History, biology, and geology begin to merge. Solving problems of excessive energy use requires coordinated work from political scientists, designers, sociologists, architects, and businessmen. As individual scholars and departments collaborate on environmental issues, the tools of enquiry are changing and becoming more widely accessible on digital platforms. Knowledge once locked inside archives and laboratories and accessible only to specialists is much more easily available. Open Access publication ensures that interdisciplinary scholarship can reach a larger audience, even as the exchange of knowledge accelerates. In this rapidly evolving context, scholars may debate to what extent the environmental humanities should collaborate with the natural or social sciences, but a generation from now the landscape of academia will likely be far different.

Traditional academic departments are themselves a product of the Anthropocene. They represent the specialization and division of knowledge into ever-smaller areas. Such specialization has not been an entirely good thing. As the polymath and early environmentalist Lewis Mumford once observed, "The key to exercising arbitrary power is to restrict the communications of individuals and groups by subdividing information so that only a small part of the whole truth will be known to any single person."[1]

In contrast, the environmental humanities are an interdisciplinary pursuit to recover endangered knowledge of environments pushed to their ragged edge and to imagine and narrate less damaging ways to be human in a shared, living world. They are not an expression of the triumphalist conquest of nature and the division of knowledge but reflect the discovery that excessive specialization has accompanied species extinction, pollution, global warming, and other human-driven interventions that collectively threaten the biosphere. The environmental humanities are part of a larger rediscovery that the world needs to be understood neither through Cartesian dualism nor as isolated fragments nor as interchangeable parts but as a vast ensemble. The university of the future may well be based on the principle that research must be interdisciplinary, in which case the departmentalization of knowledge will weaken or disappear. The fault lines of present academia may be harbingers of a fundamental realignment. Princeton University has an Environmental Institute with participation from 26 disciplines from all parts of its faculty, based on the recognition that only an interdisciplinary approach can come to grips with issues such as making cities sustainable or transforming the energy system of industrial societies.[2]

A reorganization of knowledge will not go unopposed; it will not be without costs, and it will reflect the political, economic, and cultural structures in which higher education is embedded. Within universities the departmental structure has been reified into an idealized order. Many faculty will resist a shift to a looser and more interdisciplinary formation, particularly if initiatives to fund the environmental humanities serve as a Trojan Horse to downsize traditional faculties, end lines of tenurable appointment, or cut programs deemed unpopular or "uneconomical." Countries with primarily public-funded research universities as well as those with public-private competitive systems will be tempted to fold the development of a program in the environmental humanities into a "greenwashing" of business as usual. Moreover, corporate sponsors of research may refuse to pay for philosophers, literary scholars, anthropologists, or historians as part of research teams in the natural sciences, or seek to relegate the humanities to the role of transcribers or communicators of scientific findings. Humanists might be seen as obstructions in the pipeline from academic research to corporate product development. Yet the benefits of the environmental humanities to society will become evident as environmental problems worsen and science alone proves unable to solve them. Addressing the social-environmental challenges of this century requires an interdisciplinary approach that recognizes human actions as part of the problem; a recognition that no solution will work for all locations, but rather that there are a range of possible responses that must be calibrated against local circumstances; and an understanding that multi-generational wisdom is needed for cultural transformation. If the environmental humanities teach nothing else, they teach us to mistrust quick, "total planet" solutions that promise a "green" utopia.

Opportunities

Another function of the environmental humanities will be to develop and disseminate ecologically restorative and socially just action. Some will emanate from universities. When Lisa Heller was a debate coach at the University of Richmond, she discovered that at the end of the school year the students were throwing away large quantities of perfectly good things, including bicycles, beds, radios, lamps, clothing, and anything else that might be found in a dormitory room. They had nowhere to store it, and much of it was too big to stuff in a car trunk or take home. Heller had a degree in communications with an MA thesis on the environmental activist organization Earth First!, and she could not stand to see so much waste. So she started a recycling organization called "Dump and Run," which gave students somewhere to donate unwanted things. Most of it could be sold inexpensively to incoming students in the fall. The volunteer program paid for itself, reduced waste, saved new students money, and helped instill environmental awareness in the college community. Heller provided practical advice to more than forty colleges and universities that developed similar programs. Cornell continues to run its program and generates between $40,000 and $60,000 every year that is distributed to community organizations. This example could be multiplied thousands of times, as humanities graduates identify a social-environmental problem, apply critical thinking, and take action to make a difference.

Such success stories are not rare, nor are they the special reserve of educational institutions. Another important task for the environmental humanities will be to understand new movements on behalf of restoring ecosystems that cross social status groups and defy legacies of economic exploitation or environmental racism. Rivers in western Canada have been cleaned up, and again have vigorous salmon runs, thanks to a network of local fishermen,

First Nations, and government resource managers. Species once near extinction, such as the sea otter or the American bison, have been protected and recovered through similar alliances. And successful protective actions have taken place at the planetary scale. The scientific discovery that CFCs and other chemicals were destroying the ozone layer shocked governments into action, and CFCs were banned in most of the world. While the resulting Montreal Protocol is often cited as a model for "solving" global warming, many commentators forget that CFCs were not banned without a fight. The problem was discovered in the early 1970s and confirmed by further research. As was later the case with the deniers of global warming, industries created organizations with innocuous names to fight the replacement of CFCs with alternatives. A right-wing Canadian organization warned that millions of children would die because vaccines could no longer be properly refrigerated, and there were dire predictions of food spoilage, job losses, and unnecessary sacrifices.[3] But governments listened to the scientific evidence, signed international agreements, lived up to them, and stopped the growth of the holes in the ozone layer over the Arctic and Antarctic. These holes then began to shrink. In 1995, Paul Crutzen, who had discovered that nitrogen oxides cause ozone depletion, shared a Nobel Prize for his work. Five years later he helped to coin the term "the Anthropocene."

Some corporations are changing their behavior, notably the leading cloud computing companies: Amazon, Microsoft, Facebook, Google, and Apple. Each plans to use more energy from renewables. These decisions were in part a response to negative publicity, after journalists investigated Google, Amazon, and Apple, and analyzed their energy footprint, commodity chains, e-waste streams, and work environments. One important task for graduates in the environmental humanities is to apply the skills of historical, geographical, and rhetorical analysis to pierce the veil of corporate branding. The results in the IT sector are encouraging. Both Facebook and Microsoft have the goal of 50 percent "green" energy to run their data centers by 2018, and Microsoft had already reached 44 percent in 2016. Amazon is building wind farms large enough to power 150,000 homes. But Apple outdid the competition, as in 2015 it declared that all of its data centers ran 100 percent on renewable energy. In May 2016, these and many other corporations formed the Renewable Energy Buyers Alliance, partnering with four environmental NGOs, the Rocky Mountain Institute, Business for Social Responsibility, World Resources Institute, and the World Wildlife

Fund. Their ambitious goal is to increase US alternative energy capacity by 60 gigawatts by 2025, which is equivalent to the capacity of all the old coal-fired power plants that will close by then.

Such corporate investments in "green" energy are not yet representative of government or business as a whole, however. A striking example with special resonance for the environmental humanities is the transformation of Ohio Key, an island in the Florida Keys, where Rachel Carson did much of her research for *The Edge of the Sea* (1955). By the early 1970s, however, Ohio Key had been dredged, filled, and turned into a trailer park and RV resort. (See figure 8.1.) It might have been an inspirational site of sustainable tourism, educating the public about Carson's important contributions to understanding the place of humanity and nature. But government planners and the local tourism industry erased that possibility and made Ohio Key a tourist site that treats the land as raw material and assumes traditional energy use in mobile homes, motor boats, and appliances. Such erasures are part of a larger fossil-fuel regime that connects tourism, consumption, production, and energy extraction.

Corporate lobbies for coal, oil, and gas defend the fossil-fuel regime and attack alternate energies. Just as oil companies resisted taking lead out of gasoline and tobacco companies denied the science that showed smoking causes cancer, the producers of pesticides, pharmaceuticals, and food additives vigorously defend their products. However, in the age of the Internet, consumers can easily form groups and reply to corporate public relations campaigns. Moreover, through food sharing, upcycling, and freegan websites, more and more people can opt out of the new goods market. Ad-jamming and Youtube mashup campaigns directed at corporate greenwashing, such as the one lampooning Shell's ambition to drill in the Arctic in 2012,[4] reflect effective use of art, music, and rhetoric. The environmental humanities can enable ecological citizens to push back against shortsighted corporations or government laxity in enforcement, by teaching qualitative analysis and how to use social networking tools. Already, social networks and the Internet are integrated into the environmental NGOs in Indonesia[5] and they are also becoming part of environmental activism in China.[6] There are also positive interventions in food consumption, as restaurants, schools, and businesses have changed their culinary practices as the result of campaigns and movements, like Slow Food, that underline the value of local sourcing of ecologically sound food.

Yet unrealized opportunities remain, as the problems of the global commons have not been solved. It cannot be said that the world is becoming more ecologically just, with more than a billion people going hungry while a surplus of agricultural production pushes some species to the brink of extinction. Often, social-environmental problems are not technical, but the result of a perfect moral storm: a lack of information combined with resistance to change, poorly managed institutions, and power inequalities. Despite the many successes, on the whole the environmental crisis is not being dealt with successfully. Rex Weyler, one of the founders of Greenpeace, put it this way: "We have more environment ministers, conferences, and 'protected areas,' but we have fewer species; we levy more carbon taxes yet produce greater emissions; we have more 'green' products yet have less green space. The most troubling trends—global warming, less species diversity, soil infertility, toxic dumps, shrinking forests, expanding deserts—are worsening."[7] Scientists have repeatedly issued warnings, but even the better politicians have taken half measures. Architects have designed houses that are energy neutral and whose cost over a period of thirty years is no more than a conventional house, if one includes both mortgage payments and energy costs. But banks seldom calculate mortgages with ecology in mind, and most consumers look only at the up-front cost of a house, not its total cost over time. Electric cars, if linked to and partially recharged by such a house, are also cheaper than their sticker price suggests. But the sale of gasoline cars continues to rise, often encouraged or even subsidized by national governments for whom heavy industry remains the measure of development. At national and international levels, the regulation of financial institutions seldom reflects much understanding of ecological limits or environmental justice.

The environmental humanities can provide new narratives and concepts that make necessary change attractive. The transformations that can enable a viable way of life in this century require a new cultural environment, not a series of individual gadgets. The public needs help to grasp new opportunities; both government and business could do more to speed the conversion to energy self-sufficiency. Every household can make consumer choices that will both save money and improve the environment, but to do so often requires closing the door on the twentieth-century notion that a good life is filled with more and more possessions. In the affluent West, a livable future may require a more communitarian ethics and politics and less free-wheeling individualism than was practiced during the era of "cheap nature." Individuals will benefit from cleaner air and water that come through collective down-shifting, which will reduce illnesses and cut society's health costs. The quieter city of pedestrians, bicycles and electric cars will enhance the quality of street life. People *can* break high-energy habits and become happier in the process. But doing so requires more than technologies. Sverker Sörlin noted a fundamental shift in EU policy in 2012 when the "Responses to Environmental and Societal Challenges for Our Unstable Earth (RESCUE)" group presented a report. It argued that "in a world where cultural values, political and religious ideas, and deep-seated human behaviors still rule the way people lead their lives, produce, and consume, the idea of environmentally relevant knowledge must change." It was time "to pay more attention to the human agents of the planetary pressure that environmental experts are masters at measuring but that they seem unable to prevent."[8] Change was far too slow based on "the market," which made a weak response to the crisis of global warming when it emerged in the 1980s and the 1990s or to the epidemic of species extinction in recent decades. Real change will spring from environmental values spurred by a sense of urgency and empowered by expanding access to knowledge. The environmental humanities can frame these values and make them appealing in narratives, music, video installations and other forms of vital sociality that can explain and motivate change.

The humanities are not simply an interpretive marketing tool for environmentally friendly technologies. They are necessary to recover a sense of nature and the non-human. They teach us to recognize the casualties of centuries of economic rapacity pursued under the sign of progress and to take up the reservoir of new ideas not yet realized. If wilderness is a bankrupt concept, the sublimity and complexity of nature remain lived realities. Human beings are not the only agents in the natural world. Barry Lopez recounts a story of a hunter on a snowmobile pursuing a wolverine in western Canada. First he sees only tracks in the snow, but then he begins to see the wolverine in the distance, always at the top of the next rise, always just out of gunshot. These encounters continue for some time, until at the top of one hill, he no longer sees the wolverine anywhere ahead. Then the animal leaps out of hiding and knocks him—Pow!—off the snowmobile. The wolverine gives him a long look, and then disappears into the brush. Similar stories of humans becoming almost-prey are told as primary scenes for anthropologists and philosophers—from Val Plumwood surviving a crocodile attack to Eduardo Kohn contemplating the jaguar.[9] These animal epiphanies appear to say: humans, recognize your fellow creatures and allow them space to exist. It is all too easy to forget the surprising intelligence and power of animal others until, shorn of our technological advantages, we stand again in equal if uneasy relations. For Plumwood, "these creatures indicate our preparedness to coexist with the otherness of the earth, and to recognize ourselves in mutual, ecological terms, as part of the food chain, eaten as well as eater." Such insights can help us to build refuges where ecological wonder and coexistence persist, as well as to configure the cities of the future.

Internal Challenges

While there are fine initiatives underway as well as increasing government action to deal with environmental crises, the environmental humanities face internal challenges. "Mapping Common Ground," a recent multi-author article in the journal *Environmental Humanities,* pointed out that interdisciplinarity often is more praised than practiced. It noted, for example, "Environmental historians often draw on the results of the natural sciences, but they rarely cite ecocritical scholarship or work in environmental philosophy. Ecocritics continue to invoke the virtues of interdisciplinary research, but the invoking has always been somewhat ritual in character and, when it comes to conducting the actual research, the execution is rather limited; as a whole, the field now risks becoming complacent and hidebound thanks to its hard-won and newfound respectability as a branch of literary studies."[10] Disciplinary training is not easily broadened or transformed, and the tendency is for literary critics or historians or philosophers to stay close to the practices of their fields. The rewards for breaking away from traditions are not as tangible as the permanent positions, prizes, or fellowships granted those working firmly within a discipline. There are more full time jobs to teach national literatures in designated time periods than those for scholars focused on environmental themes. Likewise, history departments primarily have permanent positions in particular eras and national histories but not necessarily in environmental history. Philosophy departments around the world seek candidates who can teach Plato, Aristotle, and the major figures in Western philosophy; they may not feel it imperative to hire a philosopher primarily concerned with environmental questions or an ethicist chiefly concerned with the rights of other species. Disciplinary habits and institutional pressures go far to explain why two fields with overlapping agendas, environmental history and ecocriticism, have found it difficult to collaborate.[11] One solution to this problem is to create institutes that hire teams of researchers drawn from multiple disciplines. In that case, however, the environmental humanities could become a specialty on the margins, while the traditional disciplines continue much as before.

Alternately, the environmental humanities might become more prominent in every department, while contributing to interdisciplinary teaching programs. They might be at the leading edge of transdisciplinary knowledge production, including knowledge co-produced with communities outside the university, knowledge firing on all (post-oil!) cylinders, with humanities integrated with social sciences and natural sciences. What might that look like? And how would it work? It might be in good part a new kind of "market research," focused on the conversion to degrowth. It would likely involve iterative processes that create new social groups and networks of expertise. And given the looming environmental challenges with a demographic implication—such as climate migrations—the transdisciplinary environmental humanities would, de facto, push the global North and academic West into non-linear, horizontal partnerships with the emerging academic East and global South.

"Mapping Common Ground" noted that another major problem for the humanities as a whole is the pressure to produce "bottom line" results, especially those that are dramatic and can be repackaged as breaking news stories. Yet many environmental problems are a form of slow violence that has emerged gradually in poor communities underserved by government, in places overexposed to risk and under-exposed in the media. The news gives extensive coverage to the tornado or the hurricane, but scarcely deals with slowly unfolding crises, such as drought in East Africa, the seepage of pesticides into a city's water supply, or the appropriation of "empty" land to build a highway through the Amazon basin. This imbalance cannot be rectified by making slow-motion disasters fit the conventional news cycle, but requires a rethinking of the news itself, moving away from sensationalism and the celebration of growth.

Like any dynamic academic field, the environmental humanities have not reached consensus on every issue. The fault lines are not necessarily between disciplines. For example, there is a major split between those who focus on the democratization of expertise and those allied with technocratic elites. On the one side are those working with environmental justice studies, citizen science, urban agriculture, transition towns, and community action research, while on the other side are experts working for sustainable development agencies or environmental governance organizations who speak the language of "grand challenges" and seek funding for large research programs. The environmental humanities include people on both sides, those linked to traditions of history from below (*Geschichte von unten*) and those who think in terms of top-down analysis and expertise.

One of the most promising methods of research in environmental humanities involves re-drawing the boundaries of how expertise is produced. Barbara Allen examined these tensions in her work on "cancer alley," the region between New Orleans and Baton Rouge notorious for its petrochemical plants and disturbing health statistics. Plants have routinely discharged chemicals into the air, soil, and water, and disasters such as Hurricanes Katrina and Rita have spread contamination from leaking facilities over a wider landscape. Local communities such as Mossville lacked both expertise and the legal and scientific vocabularies needed to mount a challenge to this pollution. They literally did not speak the language necessary to be heard, and they required outside help to gather data and document what many residents feared, that their ill health was linked to the surrounding chemical industry.[12] In this process, local people also broke down polite barriers of class, gender, and race. They developed local health statistics rather than those aggregated in broader geographical units such as states, which tend to hide health disparities linked to toxicity in specific locations. Scientists such as Florence Robinson and Wilma Subra chose popular epidemiological methods, where citizens gather illness records, and this created a different public meaning for science within the affected communities.[13] Such citizen-produced science is powerful in cases of environmental injustice, where there is deep mistrust of government officials and corporations.

Mossville residents lived in what Steve Lerner calls "sacrifice zones" where regulatory agencies turn a blind eye to toxic chemical pollution. In one Texas community that was constantly exposed, local residents had "280 times as much benzene in their blood and urine compared with people in the general population."[14] In the most severe cases it literally becomes impossible for people to live at the site. This occurred in Picher, Oklahoma, where half a century of intensive lead mining polluted the air and water so severely that the residents had to be bought out and moved away. The region has become an anti-landscape, a toxic environment that can no longer sustain human life. These sacrifice zones are especially common near mines and petrochemical factories. Environmental organizations have tended to ignore the poor in sacrifice zones. They remain largely white, middle class institutions that have "eschewed the small-scale, often contentious fence-line struggles in favor of mega-battles over climate change and protection of endangered species, forests, farmlands, open space, wetlands, and marine ecosystems."[15] Sacrifice zones are a moral imperative, calling for action against polluters and the politicians who support them or turn a blind eye. To honor the ideal of the free pursuit of truth, one must answer questions raised by abandoned mines, nuclear waste sites, and chemical dumps. In such locations, sets of interests intersect and recuperative work is possible, such as creating new archives of memories and of scarred landscapes. As David Pellow writes, "both race and class inequalities and chemical toxins operate and cooperate in ways that cause harm across social, spatial, and temporal boundaries," and these crossings can alter how humanists organize their research, since toxic inequalities also produce "opportunities for creative resistance among communities that might normally be quiescent and disconnected."[16] Universities, indigenous communities, and working class neighborhoods can come together in creative collaboration whether to support a research project, a documentary film, or a playwright's drama.

Even when environmental protection systems are in place, local populations may find themselves marginalized subjects administered by government agencies. It is rare to witness truly grassroots, populist governments that align people's interests with environmental protection. Yet alternative legal frameworks and ethical concepts for valuing nature have arisen in recent years—often in constellation with feminist, socialist, and indigenous political movements. In Ecuador and Bolivia, stronger legal protection for the environment has centered on institutionalizing the indigenous belief that Pachamama or "mother earth" has rights. This radical idea is slowly percolating into the international dialogue around law and ethics. Ecuador's revised Constitution of 2008 devotes a chapter to the "Rights of Nature," and formally opens its courts to plaintiffs on behalf of the earth from any country. The government of Bolivia passed a "Law of Mother Earth" in 2011 that includes recognition of nature's right to not be damaged by infrastructure projects. There has been some discussion at the international level of a Universal Declaration of the Rights of Nature, from the World People's Conference on Climate Change hosted in 2010 in Bolivia to a December 2015 forum in Geneva, as well as the formation of an International Rights of Nature Tribunal, a grassroots court which does not yet have legal authority.[17] The proliferation of such alternative world institutions reflects an increase in dissatisfaction. Global environmental policy is still dominated by Western governments and organizations that may be influenced by multinational corporations. This creates tensions at summit meetings where the different constituencies meet. Moreover, the traditional ecological knowledge of local communities is often ignored or marginalized in the legal frameworks that are intended to provide regulation and protection. The environmental humanities can play an essential role in these dialogues, particularly by bridging the gaps between critical legal studies and history, philosophy, and culture.

Can one merely study a severely polluted community without becoming an expert witness in its defense? Perhaps not. For the sacrifice zones are spreading and the "suffering of the other," Chris Hedges argues, "is universal." In *Days of Destruction, Days of Revolt* (illustrated by the graphic artist Joe Sacco) Hedges surveys sacrifice zones such as West Virginia's mountaintop strip mines and the tomato plantations of Florida, where since 2007 the US Justice Department has prosecuted seven cases of slavery. From a street-level view of the Occupy Wall Street movement in 2011, Hedges observed: "The corporate leviathan has migrated with the steady and ominous thud of destruction from the outer sacrifice zones to devour what remains. The vaunted American dream, the idea that life will get better, that progress is inevitable if we obey the rules and work hard, that material prosperity is assured, has been replaced by a hard and bitter truth. The American dream, we know, is a lie. We will all be sacrificed."[18] In his view, activism is the only choice, though this is not traditionally the role of academic researchers. Yet interventions may come in many forms, and do not necessarily mean joining protests in the street. Martin Melosi, an historian of sanitation and waste removal, has often served as an expert witness. Notably, he was employed by the US Department of Justice in a four-year trial involving the Shell Oil Corporation.

This book has examined the emergence and meaning of the environmental humanities. It has argued that the philosophical focus on "the thing in itself" should be abandoned in favor of the pragmatic study of organisms in their habitat, that human cultures also must be understood as part of a larger environmental context, that human beings do not have exceptional rights relative to other species, and that industrial Western cultures are not inherently superior. Failure to understand these ideas has led to the crisis of the commons, the destruction of companion species in the sixth extinction, ecological imperialism, and the Anthropocene. Within human communities, failure to grasp these ideas underlies environmental injustice, ecoracism and slow violence.

In response to these multiple crises, the environmental humanities early began to cultivate new ways of understanding humanity's linkages to nature. Nonfiction nature writers and cultural geographers focused on the human relationship to particular sites and "place-making." Their interest in understanding place coincided with both the determination to preserve wilderness areas and the rise of ecotourism. But the concept of wilderness proved to be flawed, and if the impulses behind ecotourism were admirable, it had contradictory effects. "Protected areas" such as national parks at times attracted so many tourists that they endangered the species needing protection, while incurring costs to local communities. In society as a whole, efforts to achieve sustainability focused on energy conservation, on making consumption less resource demanding, and on redeveloping cities. Before the reliance on cheap fossil fuels, cities sourced most of their food locally, had many gardens, and recycled much of their waste. A reinvention of these practices using alternative energy technologies is already underway and can reduce the human impact on the environment.

Alternately, some scientists argue for large-scale manipulations of the environment. Should one launch thousands of small deflectors into space to cut down the sunlight reaching the earth? Might rising seas be diverted into the Sahara desert? Is it a good idea to modify the enzymes in cow stomachs in order to reduce their methane discharge? These are potentially dangerous proposals that would add to the complexities of the Anthropocene, as would geo-engineering, the manipulation of DNA, and the possibile human co-evolution with machines into cyborgs.

Humanity has reached a decisive turning point. Its activities have had unintended consequences, destroying habitats, eliminating species, changing the chemistry of lakes and oceans, and creating the greenhouse effect of global warming. The death of other species raises the specter of human extinction, or less drastically, the possible collapse of civilizations and the survival of remnants of life in difficult circumstances. Artists, novelists, and filmmakers have considered these dire outcomes and made them a part of cultural awareness well before political parties felt the urgency of the problems and developed the will to act. Indeed, even in 2017 some politicians deny or minimize the existence of global warming, notably Donald Trump. Precisely because this resistance continues, the environmental humanities must engage dark visions of the anthropocene as part of a larger effort to understand the problems humankind faces and to imagine the new values, new forms of citizenship, and new social practices that will be necessary to carry humanity into better times. The environmental humanities help sort out the merely fanciful from the urgently possible futures, enabling the public to move past denial, anger, and negotiation to action. The imagination of disaster is not necessarily the prelude to apocalypse but rather a stimulus to avoid it. Constructive changes are being realized by communities in all parts of the world, including localization, degrowth, advanced recycling systems, and the emergence of commodity regionalism. Likewise, the mainstreaming of ecological economics is essential to environmental progress. In addition, scholars of new materialism, animal rights, and queer theory argue that other fundamental revolutions in consciousness are required, at times taking inspiration from aboriginal peoples.

The Iroquois Confederation believed that a society must think ahead for seven generations.[19] We hope there will still be students two centuries from now, and that they will look back at the early twenty-first century and see the emergence of the environmental humanities as part of a fundamental reorientation, from overspecialization in the academy to new forms of teamwork and interdisciplinarity, from bogus cultural hierarchies to cultural equality, from excessive resource extraction to biomimetic recycling, from the sixth extinction to species revival, and from unsustainable economic growth to a society based not on rapid obsolescence but on durability and respect for other species and their environments. This book seeks to speed that transition.

Notes

Chapter 1

1 Christina Larson, "China's Grand Plans for Eco Cities Now Lie Abandoned," Yale Environment 360 blog, April 6, 2009. http://www.christina-larson.com/china%e2%80%99s-grand-plans-for-eco-cities-now-lie-abandoned/

2 See *American Earth: Environmental Writing Since Thoreau*, ed. Bill McKibben (Library of America, 2008).

3 Richard Grove and Vinita Damodaran, "Imperialism, Intellectual Networks, and Environmental Change; Unearthing the Origins and Evolution of Global Environmental History," in *Nature's End: History and Environment*, ed. Sverker Sörlin and Paul Warde (Palgrave Macmillan, 2009).

4 *The Future of Nature*, ed. Libby Robin, Sverker Sörlin, and Paul Warde (Yale University Press, 2013).

5 Tom Griffiths, "The Humanities and an Environmentally Sustainable Australia," *Australian Humanities Review*, December 2007 (australianhumanitiesreview.org). The essay first appeared in a 2003 report by Australia's Department of Education titled *The Humanities and Australia's National Research Priorities*.

6 Janet Abbate, *Inventing the Internet* (MIT Press, 1999), 83–144.

7 David E. Nye, *Technology Matters: Questions to Live With* (MIT Press, 2006), 33–45.

8 See David E. Nye, *America as Second Creation* (MIT Press, 2003), 1–20, 294–297.

9 Carolyn Merchant, "Reinventing Eden," in *Uncommon Ground*, ed. William Cronon (Norton, 1996), 157–158.

10 The Rhode Island School of Design has been a major sponsor of the "STEM to STEAM" movement. For more on the movement, see http://stemtosteam.org/.

11 John Robinson, "Being Undisciplined: Transgressions and Intersections in Academia and Beyond," *Futures* 40 (2008): 74; Poul Holm et al., "Collaboration between

the Natural, Social and Human Sciences in Global Change Research," *Environmental Science and Policy* 28 (2013): 26.

12 Libby Robin, "New Science for Sustainability in an Ancient Land," in *Nature's End: History and the Environment*, ed. Sverker Sörlin and Paul Warde (Palgrave Mac-Millan, 2009), 188.

13 The spring 1998 special issue of the journal *Telos*, edited by David Pan and dedicated to the "Crisis in Higher Education," offers an overview of how American liberals, leftists, and conservatives struggled over the purpose and meaning of humanities training. Discussion elsewhere centered on the politicization of research, the job market crisis, and the number and size of graduate programs. Similar debates over the "crisis of the humanities" occurred in subfields and worldwide, from feminist studies in medieval literature to the teaching of African languages in post-Apartheid South Africa.

14 An international and interdisciplinary sample in this spirit includes the following works: Griffiths, "The Humanities and an Environmentally Sustainable Australia" (cited above); Ursula Heise, "The Hitchhiker's Guide to Ecocriticism," *PMLA* 121, no. 2 (2006): 503–516; Jody A. Roberts and Nancy Langston, "Toxic Bodies/ Toxic Environments: An Interdisciplinary Forum," *Environmental History* 13, no. 4 (2008): 629–635.

15 On this complex topic, see Ruediger H. Grimm, *Nietzsche's Theory of Knowledge* (Walter de Gruyter, 1977), 56–57. See also Jane Bennett, *Vibrant Matter* (Duke University Press, 2010).

16 Alan Watts, "The World Is Your Body," in *The Ecological Conscience*, ed. Robert Disch (Prentice-Hall, 1970), 182.

17 Raymond Williams, "Ideas of Nature," in *Problems in Materialism and Culture* (Verso, 1980).

18 Jeffrey Goldberg, "Were There Dinosaurs on Noah's Ark?" *The Atlantic*, October 2014.

19 Kathleen Dean Moore and Michael P. Nelson, eds., *Moral Ground: Ethical Action for a Planet in Peril* (Trinity University Press, 2011).

20 http://w2.vatican.va/content/francesco/en/encyclicals/documents/papa-francesco_20150524_enciclica-laudato-si.html

21 Ray Kurzweil, *The Singularity Is Near* (Viking, 2005).

22 Kevin Kelly, *What Technology Wants* (Viking, 2010), 358.

23 Thom van Dooren, *Flight Ways: Life and Loss at the Edge of Extinction* (Columbia University Press, 2014).

24 John Neihardt, ed., *Black Elk Speaks* (University of Nebraska Press, 1979).

25 Michael Adas, *Machines as the Measure of Men: Science, Technology, and Ideologies of Western Dominance* (Cornell University Press, 1989).

26 Derek Wall, *The Commons in History: Culture, Conflict and Ecology* (MIT Press, 2014), 25–28.

27 Hardin's article in *Science*, which was based on a legal case study of California's fisheries by Arthur F. McEvoy, was seized upon as a justification for global financial mechanisms that would assign private owners to previously public goods, particularly in developing countries. For a nuanced reading of Hardin, see Matthew MacLellan, "The Tragedy of Limitless Growth: Re-interpreting the Tragedy of the Commons for a Century of Climate Change," *Environmental Humanities* 7 (2015): 41–58.

28 Wall, *The Commons in History*, 134. Also see https://www.eff.org/about.

29 https://www.eff.org/about

30 Nancy Cowan, *Peregrine Spring* (Rowman & Littlefield, 2016), 254.

31 Donna Haraway, *When Species Meet* (University of Minnesota Press, 2008).

32 Robert Traer, *Doing Environmental Ethics* (Westview, 2009), 150.

33 Elizabeth Kolbert, *The Sixth Extinction* (Bloomsbury Books, 2014).

34 Peter Matthiessen's classic book *Wildlife of America* (Viking, 1959) opens with a moving description of this encounter and the fate of the last pair of Great Auks.

35 Kolbert, *The Sixth Extinction,* 141.

36 Alfred W. Crosby, *Ecological Imperialism*, second edition (Cambridge University Press, 2015).

37 Mike Hulme, ed., *Climates and Cultures* (SAGE, 2015).

38 Martin Melosi, "Equity, Eco-racism and Environmental History," *Environmental History Review* 19 (fall 1995): 4. See also Robert D. Bullard, *The Quest for Environmental Justice: Human Rights and the Politics of Pollution* (Counterpoint, 2005).

39 Barbara Allen, "Narrating the Toxic Landscape of 'Cancer Alley,' Louisiana," in *Technologies of Landscape: Reaping to Recycling*, ed. David E. Nye (University of Massachusetts Press, 1999); Allen, *Uneasy Alchemy: Citizens and Experts in Louisiana's Chemical Corridor Disputes* (MIT Press, 2003), especially pp. 1–13; Joseph Pratt and Martin Melosi, "The Energy Capital of the World? Oil-Led Development in Twentieth Century Houston," in *Energy Capitals: Local Impact, Global Influence*, ed. Pratt and Melosi (University of Pittsburgh Press, 2014), 52–56.

40 Melosi, "Equity, Eco-racism and Environmental History," 4.

41 Joni Adamson and Slovic Scott, "The Shoulders We Stand On: An Introduction to Ethnicity and Ecocriticism," *MELUS* 34, no. 2 (2009): 20.

42 Timothy Doyle, *Environmental Movements* (Rutgers University Press, 2005), 60–67.

43 David N. Pellow, *Resisting Global Toxics: Transnational Movements for Environmental Justice* (MIT Press, 2007).

44 Julie Schmit, "Report: U.S. Lax on Exports of Toxic E-Waste from Old E-Gear," *USA Today*, September 18, 2008.

45 Rob Nixon, *Slow Violence and the Environmentalism of the Poor* (Harvard University Press, 2011).

46 Michael Buckley, *Meltdown in Tibet: China's Reckless Destruction of the Ecosystems from the Highlands of Tibet to the Deltas of Asia* (Palgrave Macmillan, 2014).

47 Amitav Ghosh, *The Hungry Tide* (Houghton Mifflin Harcourt, 2005).

48 C. S. Holling, "Resilience and Stability of Ecological Systems," *Annual Review of Ecology and Systematics* 4 (1973): 1–23.

49 See *Resilience: A Journal of the Environmental Humanities* 1, no. 1 (2014).

50 Editors' Column, ibid., 4.

Chapter 2

1 Mitchell Thomashow, *Ecological Identity* (MIT Press, 1995), 9.

2 Klara Bonsack Kelley and Harris Francis, *Navajo Sacred Places* (Indiana University Press, 1994), 38–39.

3 Yi-Fu Tuan, *Space and Place: The Perspective of Experience* (University of Minnesota Press, 1977), 164.

4 Deborah Tall, *From Where We Stand: Recovering a Sense of Place* (Johns Hopkins University Press, 1993).

5 See Nye, *America as Second Creation*, 9–42.

6 Jeremy Kendal et al. "Human Niche Construction in Interdisciplinary Focus," *Philosophical Transactions of the Royal Society B* 366 (2011): 785-792.

7 Rachel Carson, *Silent Spring* (Houghton Mifflin, 1962).

8 George Perkins Marsh, *Man and Nature* (University of Washington Press, 2003), 9.

9 Barry Lopez, "Landscape and Narrative," in Lopez, *Crossing Open Ground* (Vintage Books, 1989), 67–68.

10 Ibid., 71.

11 Tuan, *Space and Place*, 170.

12 We are indebted to an anonymous reviewer for pointing out the importance of these differences.

13 Madhav Gadgil and Ramachandra Guha, *This Fissured Land: An Ecological History of India* (University of California Press, 1993).

14 James David Fahn, *A Land on Fire: The Environmental Consequences of the Southeast Asian Boom* (Westview, 2003), 7.

15 Ursula K. Heise, *Sense of Place, Sense of Planet: The Environmental Imagination of the Global* (Oxford University Press, 2008), 210.

16 Richard Grove, *Green Imperialism: Colonial Expansion, Tropical Island Edens and the Origins of Environmentalism, 1600–1860* (Cambridge University Press, 1996).

17 Dean MacCannell, *The Tourist* (Schocken, 1989), 1.

18 Don DeLillo, *White Noise* (Penguin Classics, 2009).

19 MacCannell, *The Tourist*, 142–143.

20 John Urry, *The Tourist Gaze* (SAGE, 1990), 64.

21 James Buzard, *The Beaten Track* (Clarendon, 1993).

22 See also https://conservationvip.org/.

23 Zoë Meletis and Lisa Campbell, "Benevolent and Benign? Using Environmental Justice to Investigate Waste-Related Impacts of Ecotourism in Destination Communities," *Antipode* 41, no. 4 (2009): 760.

24 Erving Goffman, *The Presentation of Self in Everyday Life* (Anchor Books, 1959).

25 Anthony Carrigan, "Out of This Great Tragedy Will Come a World Class Tourism Destination," in *Postcolonial Ecologies: Literatures of the Environment*, ed. Elizabeth DeLoughrey and George B. Handley (Oxford University Press, 2011), 276.

26 *Jikoo / La Chose Espérée*, dir. Christophe Leroy and Adrien Camus (La Troisieme Porte a Gauche films, 2014).

27 http://www.unwto.org/step/about/en/step.php?op=1

28 Joachim Radkau, *Nature and Power: A Global History of the Environment* (Cambridge University Press, 2008), 280–282.

29 Cecilia Alvear, "Are We Loving the Galápagos to Death?" *Harvard Review of Latin America* 8, no. 3 (2009): 22–23; Robin M. Self, Donald Self, and Janel Bell-Haynes, "Marketing Tourism In The Galapagos Islands: Ecotourism or Greenwashing?" *International Business and Economics Research Journal* 9, no. 6 (2010): 111–125.

30 Radkau, *Nature and Power*, 284–285.

31 Christina Eisenberg, *The Carnivore Way: Coexisting with and Conserving North America's Predators* (Island, 2014), 63.

32 Rolf O. Peterson and John A. Vucetich, Ecological Studies of Wolves on Isle Royale: Annual Report 2015–16, 2.

33 See www.isleroyalewolf.org or visit the Facebook page of the wolves of Isle Royale.

34 Emma Marris, *Rambunctious Garden: Saving Nature in a Post-Wild World* (Bloomsbury, 2011).

35 Eben Kirksey, *Emergent Ecologies* (Duke University Press, 2015).

36 Eduardo Kohn, *How Forests Think: Toward an Anthropology Beyond the Human* (University of California Press, 2013).

37 On the rewilding of upstate New York, see Bill McKibben, *Hope, Human and Wild: True Stories of Living Lightly on the Earth* (Milkweed: 2007), chapter 1. On the return of species to Scandinavia, see http://dolly.jorgensenweb.net/nordicnature; Dolly Jørgensen, "Migrant Muskoxen and the Naturalization of National Identity in Scandinavia," in *The Historical Animal*, ed. Susan Nance (Syracuse University Press, 2015).

38 For an overview of the Oostvaardersplassen in English, see Elizabeth Kolbert, "Recall of the Wild: The Quest to Engineer a World Before Humans," *The New Yorker*, December 24–31, 2012.

39 The popularity of Verkerk's 2013 documentary *De Nieuwe Wildernis* led to a sequel, *Holland: Natuur in de Delta* (2015); together they represent a strand of conservationist nationalism and rebranding of the Netherlands as an ecologically friendly country.

40 Elizabeth DeLoughrey and George B. Handley, "Introduction: Toward an Aesthetics of the Earth," in *Postcolonial Ecologies: Literatures of the Environment*, ed. Elizabeth DeLoughrey and George B. Handley (Oxford University Press, 2011), 20–24.

41 For an excellent range of international perspectives on wilderness in relation to empire, see *Ecology and Empire: Environmental History of Settler Societies*, ed. Tom Griffiths and Libby Robin (University of Washington Press, 1997). For an example of how the categories of nature and culture prove hard to separate, see Stephen J. Pyne, *Fire: A Brief History* (University of Washington Press, 2001), 1–25; Jane Carruthers, "Nationhood and National Parks: Comparative Examples from the Post-imperial Experience," in *Ecology and Empire: Environmental History of Settler Societies*, ed. Tom Griffiths and Libby Robin (University of Washington Press, 1997), 133.

42 William Cronon, "The Trouble with Wilderness: Or, Getting Back to the Wrong Nature," in *Uncommon Ground: Rethinking the Human Place in Nature*, ed. Cronon (Norton, 1995).

43 The language of the Wilderness Act (often credited to Howard Zahnisser, who was president of the Wilderness Society when the act was written) borrows directly from the Wilderness Society's founder, Bob Marshall, who defined wilderness in precisely these terms in "The Problem of the Wilderness" (*Scientific Monthly* 2, February 1930, 141–148).

44 Madhav Gadgil and Ramachandra Guha, *The Use and Abuse of Nature* (Oxford University Press, 2000), 148–150.

45 Karl Jacoby, *Crimes Against Nature: Squatters, Poachers, Thieves, and the Hidden History of American Conservation* (University of California Press, 2014). Conflicts between the cattle kept by Maasai pastoralists and black rhino and conflicts between farmers and elephants in the Serengeti and Mara reserves in Kenya and Tanzania are paradigmatic. Studies of pressures from local subsistence farming and hunting versus those of tourist activities that involve local stakeholders are particularly important; for one example, see M. J. Walpole, G. G. Karanja, N. W. Sitati, and N. Leader-Williams, "Wildlife and People: Conflict and Conservation in Masai Mara, Kenya," Wildlife and Development Series no. 14, International Institute for Environment and Development, London.

46 Jane Carruthers, *The Kruger National Park: A Social and Political History* (University of Natal Press, 1995); Christopher Conte, *Highland Sanctuary: Environment and History in Tanzania's Usambara Mountains* (Ohio University Press, 2004), 68, 85.

47 "Wilderness Babel" is permanently hosted by the Environment & Society Digital Portal: http://www.environmentandsociety.org/exhibitions/wilderness/

48 This is a point of consensus among indigenous environmental groups in North America, the Indigenous Environmental Network (founded in 1990 by the Navajo in Arizona), and the First Nations Environmental Network (formed in Canada in 1992).

49 Lars Elenius, "Vildmark and ödemark—Swedish," Environment & Society Portal Virtual Exhibitions, Environmentandsociety.org.

50 Source: Kalevi Kull, biosemiotics forest walking lecture given during "Animals in Transdisciplinary Environmental History" ESEH Summer school, Haapsalu, Estonia, 2015. Also see Kadri Tüür's entry on "Metsik Loodus, Puutumatu Loodus and Põlisloodus—Estonian" in the "Wilderness Babel" project.

51 On the complex cultural and environmental history that transformed the Japanese wolf from a venerated creature to one hunted into extinction, see Brett L. Walker, *The Lost Wolves of Japan* (University of Washington Press, 2005).

52 Natasha Yamamoto, "Defining Wilderness—Japanese" in "Wilderness Babel" virtual exhibition.

53 Sebastião Salgado, *Genesis* (Taschen, 2013), 7.

54 Ibid., 6.

55 Ibid., 464, 469. The two images mentioned are full-page photographs and fold inward so as to appear opposite one another in the book, like early modern Christian diptychs of Adam and Eve.

56 Rupert Stasch, "Toward Symmetric Treatment of Imaginaries: Nudity and Payment in Tourism to Papua's 'Treehouse People,'"" in *Tourism Imaginaries: Anthropological Approaches*, ed. Noel B. Salazar and Nelson H. H. Graban (Berghahn, 2014), 38–40.

57 Gary Snyder, *The Practice of the Wild* (Milkweed, 1990), 15.

58 Iain McCalman, *The Reef: A Passionate History* (Farrar, Straus and Giroux, 2013), 10.

Chapter 3

1 http://www.un.org/sustainabledevelopment/cities/

2 Leslie White, *Science of Culture* (Grove, 1949), 369. For a different view of energy history, see David E. Nye, *Consuming Power* (MIT Press, 1998).

3 Timothy Mitchell, *Carbon Democracy: Political Power in the Age of Oil* (Verso, 2011), 229; Nye, *Consuming Power*, 217–223.

4 Harvey Blatt, *America's Environmental Report Card* (MIT Press, 2005), 210.

5 Ibid., 216.

6 David E. Nye, *When the Lights Went Out* (MIT Press, 2010), 212–213.

7 John McNeill, *Something New Under the Sun: An Environmental History of the Twentieth Century World* (Norton, 2000), 311.

8 Frank Trentmann, *Empire of Things: How We Became a World of Consumers, from the Fifteenth Century to the Twenty-First* (Penguin, 2016), 671.

9 Benjamin Sovacool, *The Dirty Energy Dilemma: What's Blocking Clean Power in the United States* (Praeger, 2008), 80–81.

10 John A. Laitner and Karen Ehrhardt-Martinez, "Information and Communication Technologies: The Power of Productivity," report E081, American Council for an Energy Efficient Economy, 2008, 21, 26.

11 John M. Deutch, *The Crisis in Energy Policy* (Harvard University Press, 2011), 37.

12 Arthur Neslen, "Portugal Runs for Four Days Straight on Renewable Energy Alone," *The Guardian*, May 18, 2016.

13 Quoted in Carolyn Lochhead, "How GOP Became Party of Denial on Global Warming," *San Francisco Chronicle*, April 28, 2013.

14 On the early history of consumption, see Ruth Schwartz Cowan, *More Work for Mother* (Basic Books, 1983); Susan Strasser, *Satisfaction Guaranteed: The Making of the American Mass Market* (Pantheon, 1989).

15 The classic critique of the supposed power of advertising is Michael Schudsen, *Advertising: The Uneasy Persuasion* (Basic Books, 1984).

16 Lisbeth Cohen, "Encountering Mass Culture at the Grassroots: The Experience of Chicago Workers in the 1920s," *American Quarterly* 41, no. 1 (1989): 6–33.

17 Lisbeth Cohen, *A Consumer's Republic* (Knopf, 2003).

18 Marion Nestle, *Food Politics: How the Food Industry Influences Nutrition and Health*, revised edition (University of California Press, 2007).

19 Warren J. Belasco, *Appetite for Change: How the Counterculture Took On the Food Industry* (Cornell University Press, 1993).

20 http://www.footprintnetwork.org/ecological_footprint_nations/ecological_per_capita.html

21 Njoki Nathani Wane, *Indigenous African Knowledge Production: Food-Processing Practices among Kenyan Rural Women* (University of Toronto Press, 2014), 22.

22 "Heat-Trapping Gas Passes Milestone, Raising Fears," *New York Times*, May 13, 2013.

23 "Globe 'Woefully Unprepared' for Rise in Disasters Linked to Climate Change," *The Guardian*, May 17, 2016.

24 Isabelle Stengers, *Au temps des catastrophes: Résister à la barbarie qui vient* (Découverte, 2013), 11 (translation by Robert Emmett).

25 Ibid., 22.

26 Tom Princen and Michael Maniates, *Confronting Consumption* (MIT Press, 2002), 5.

27 Wendell Berry, *The Unsettling of America* (Sierra Club, 1977).

28 Thomas Merton, *Conjectures of a Guilty Bystander* (Doubleday, 1966), 222.

29 Letter from Thomas Merton to Rachel Carson, January 13, 1963, reproduced in Monica Weiss, *The Environmental Vision of Thomas Merton* (University Press of Kentucky, 2011), 14–15.

30 Jens Lachmund, *Greening Berlin: The Co-production of Science, Politics, and Urban Nature* (MIT Press, 2013).

31 Jeffrey Hou, *Insurgent Public Space: Guerrilla Urbanism and the Remaking of Contemporary Cities* (Routledge, 2010).

32 http://www.un.org/sustainabledevelopment/cities/

33 Ted Steinberg, "The Death of the Organic City," in *Down to Earth: Nature's Role in American History* (Oxford University Press, 2002), 157–172.

34 http://www.vaxjo.se/sustainable

35 David Owen, *Green Metropolis* (Riverhead Books, 2009).

36 On the cultural roots and politics of urban agriculture in the United States, see Robert S. Emmett, *Cultivating Environmental Justice* (University of Massachusetts Press, 2016), 132–167. On the resurgence of backyard gardening in Australia, see Andrea Gaynor, *Harvest of the Suburbs: An Environmental History of Growing Food in Australian Cities* (University of Western Australia Press, 2006).

37 Owen, *Green Metropolis*, 303.

38 Duncan McLaren and Julian Agyeman, *Sharing Cities: A Case for Truly Smart and Sustainable Cities* (MIT Press, 2015), 231–234.

39 Citied in ibid., 314.

40 Stephen Moore, *Alternative Routes to the Sustainable City* (Lexington Books, 2007), 62–63.

41 Ibid., 148–149.

42 Erica Goode, "New Solar Plants Generate Floating Green Power," *New York Times*, May 20, 2016.

43 The "cycling cities" research program based in Eindhoven, the Netherlands includes several books, a new media platform, and planned exhibitions with partners across Asia, Europe, and North America. See http://www.cyclingcities.info/research/cpsum/.

Chapter 4

1 The bioethicist Jonathan D. Moreno reviewed the state of the art of germ-line gene editing and framed the many resulting ethical issues in an online *Scientific American* article titled "Where to Draw the Line on Gene-Editing Technology" (https://www.scientificamerican.com/article/where-to-draw-the-line-on-gene-editing-technology, November 30, 2015).

2 http://www.scimagojr.com/journalrank.php?area=1300

3 Antonio Regalado, "Paint-On GMOs Could Create Cattle, Dogs, with Custom Fur," *MIT Technology Review*, May 17, 2016. https://www.technologyreview.com/s/601448/paint-on-gmos-could-create-cattle-dogs-with-custom-fur/

4 The BRAIN initiative, launched in 2013, and held its third annual meeting for funded researchers in December 2016.

5 Antonio Regalado, "Ethical Questions Loom Over Efforts to Make a Human Genome from Scratch," *MIT Technology Review*, May 25, 2016. https://www. technologyreview.com/s/601540/ethical-questions-loom-over-efforts-to-make-a -human-genome-from-scratch/

6 Ibid.

7 "What If a Species Could Be Brought Back?" *New York Times*, March 20, 2013.

8 Natasha Meyers, *Rendering Life Molecular: Models, Modelers, and Excitable Matter* (Duke University Press, 2015), 25.

9 Ibid.

10 We are grateful to the anthropologist Filippo Bertoni for introducing us to this work. Bertoni is currently exploring the implications of astrobiology for space mining in the context of mega-mine projects such as those managed by the Rio Tinto Group.

11 Meyers, *Rendering Life Molecular*, 26.

12 Ibid., 27.

13 Ibid., xi.

14 Ibid., 237.

15 Vandana Shiva, *Biopiracy: The Plunder of Nature and Knowledge* (South End, 1997).

16 "One Per Cent: Grow your own living lights," *New Scientist*, May 4, 2013.https:// www.newscientist.com/article/mg21829156-500-one-per-cent-grow-your-own -living-lights/

17 Lynn Margulis, *Symbiotic Planet: A New Look at Evolution* (Basic Books, 1998), 3.

18 North American Bird Conservation Initiative, The State of North America's Birds 2016. http://www.stateofthebirds.org

19 Michael Ohl, *Die Kunst der Benennung* (Matthes & Seitz, 2015), 7, 177.

20 Mark Barrow, *Nature's Ghosts: Confronting Extinction from the Age of Jefferson to the Age of Ecology* (University of Chicago Press, 2009), 348 (emphasis added).

21 "History of the Convention on Biological Diversity," https://www.cbd.int/ history/

22 Ernst Mayr, *This Is Biology: The Science of the Living World* (Harvard University Press, 1998), 37.

23 United Nations Development Program, "Indonesia Launches National Blueprint to Protect its Biodiversity," January 21, 2016. http://www.id.undp.org/content/

indonesia/en/home/presscenter/pressreleases/2016/01/21/indonesia-launches
-national-blueprint-to-protect-its-biodiversity.html

24 Cheryl Lousley, "E. O. Wilson's Biodiversity, Commodity Culture, and Sentimental Globalism," *RCC Perspectives* 2012, no. 9: 11–16.

25 Henry David Thoreau, *A Week on the Concord and Merrimack Rivers* (Penguin,
1998), 28–29.

26 Ibid., 31.

27 Thom van Dooren, *Flight Ways: Life and Loss at the Edge of Extinction* (Columbia
University Press, 2014), 58.

28 Chris D. Thomas et al., "Extinction Rate from Climate Change," *Nature* 427
(January 8, 2004): 145–148.

29 Eben Kirksey and and Stefan Helmreich, "The Emergence of Multispecies Ethnography: A Special Guest-Edited Issue of Cultural Anthropology," *Cultural Anthropology* 25, no. 4 (2010): 545.

30 Donna Haraway, *When Species Meet* (University of Minnesota Press, 2008), 244.

31 van Dooren, *Flight Ways*, 117.

32 Mauro Agnoletti, "Rural Landscape, Nature Conservation and Culture: Some
Notes on Research Trends and Management Approaches from a (Southern) European Perspective," *Landscape and Urban Planning* 126 (2014): 72.

33 Philip P. Micklin, "Desiccation of the Aral Sea: A Water Management Disaster in
the Soviet Union," *Science* 241 (September 2, 1988): 1170–1176.

34 Paul Crutzen, "Albedo Enhancement by Stratospheric Sulfur Injections: A Contribution to Resolve a Policy Dilemma?" *Climatic Change*, August 2006, 211.

35 Alexander M. Stoner and Andony Melathopoulos, *Freedom in the Anthropocene:
Twentieth-Century Helplessness in the Face of Climate Change* (Palgrave, 2015), 14.

36 James Rodger Fleming, *Fixing the Sky: The Checkered History of Weather and Climate Control* (Columbia University Press, 2010), 179.

37 Clive Hamilton, *Earthmasters: The Dawn of the Age of Climate Engineering* (Yale
University Press, 2013), 154–156.

38 Ibid., x.

39 Jedediah Purdy, "Anthropocene Fever," *Aeon*, March 31, 2015. https://aeon.co/
essays/should-we-be-suspicious-of-the-anthropocene-idea

40 James Fleming, "Climate Engineering is Untested and Dangerous," NewSecurity-
Beat, August 20, 2009. https://www.newsecuritybeat.org/2009/08/climate-engineering
-is-untested-and-dangerous/

41 Stephen M. Gardiner, "Some Early Ethics of Geoengineering the Climate: A Commentary on the Values of the Royal Society Report," *Environmental Values* 20 (2011): 163–188.

42 Ibid., 167.

43 See Sobecka's "Cloud Machine" (http://www.amateurhuman.org/projects/cloud -maker).

44 Naomi Oreskes and Erik Conway, *The Collapse of Western Civilization: A View from the Future* (Columbia University Press, 2014), 32.

45 Andrew Pollack, "Scientists Talk Privately About Creating a Synthetic Human Genome," *New York Times*, May 13, 2016.

46 Aldous Huxley, *Brave New World* (Chatto and Windus, 1932).

47 Donna Haraway, "A Manifesto for Cyborgs," in *The Haraway Reader* (Routledge, 2004).

48 Katharine Hayles, *How We Became Posthuman: Virtual Bodies in Cybernetics, Literature, and Informatics* (University of Chicago Press, 1999).

49 "An interview/dialogue with Albert Borgmann and N. Katherine Hayles on humans and machines," http://press.uchicago.edu/Misc/Chicago/borghayl.html

Chapter 5

1 Aldo Leopold, "The Round River," in *A Sand County Almanac with Essays on Conservation from Round River* (Ballantine, 1970), 197.

2 These are delivered in the careful, calm language of each five-year assessment report which synthesizes the findings of thousands of independent researchers. All are available online at the IPCC's publication hub, for example: http://www.ipcc.ch/ report/ar5/wg2/docs/WGIIAR5_SPM_Top_Level_Findings.pdf

3 To sample this genre, see *The End of Doom* by Ronald Bailey, from the Cato Institute. On industry-funded attacks on environmentalists and scientists see Frederick Buell, *From Apocalypse to Way of Life* (Routledge, 2003), 3–37. On industry-funded scientists denying evidence linking products to cancers and global warming, see Naomi Oreskes and Erik Conway, *Merchants of Doubt* (Bloomsbury, 2010) and the 2014 Robert Kenner film of the same name.

4 In addition to the Worldwatch Institute's annual State of the World reports, a vast body of research-based popular nonfiction documents the ecological crisis. See, e.g., Peter F. Sale, *Our Dying Planet: An Ecologist's View of the Crisis We Face* (University of California Press, 2011); P. H. Liotta and Allan W. Shearer, *Gaia's Revenge: Climate Change and Humanity's Loss* (Praeger, 2007).

5 Roy Scranton, *Learning to Die in the Anthropocene: Reflections on the End of Civilization* (City Lights, 2015), 19.

6 See Joanna Zylinska, *Minimal Ethics of the Anthropocene* (Open Humanities Press, 2014); Jedediah Purdy, *After Nature: A Politics for the Anthropocene* (Harvard University Press, 2015).

7 Paul Crutzen and Eugene Stoermer, "The 'Anthropocene,'" *IGBP Newsletter* 41 (May 2000): 18.

8 Gregg Mitman, "Hubris or Humility: Genealogies of the Anthropocene," in *Future Remains: A Cabinet of Curiosities for the Anthropocene*, ed. Gregg Mitman, Marco Armiero, and Robert S. Emmett (University of Chicago Press, 2017).

9 Jan Zalasiewicz et al., "The Technofossil Record of Humans," *Anthropocene Review* 1, no. 1 (2014): 34.

10 Jan Zalasiewicz, "Is Earth in a New Geological Phase Thanks to Us?" *New Scientist*, November 5, 2014. https://www.newscientist.com/article/mg22429940-200-is-earth-in-a-new-geological-phase-thanks-to-us/

11 Lori A. Ziolkowsi, "The Geologic Challenge of the Anthropocene," *RCC Perspectives* 2016, no. 2: 38.

12 Andrey Ganopolski et al., "Critical Insolation–CO_2 Relation for Diagnosing Past and Future Glacial Inception," *Nature* 529 (January 14, 2016): 203.

13 Dipesh Chakrabarty, "The Climate of History: Four Theses," *Critical Inquiry* 35, no. 2 (2009): 201.

14 Shawn William Miller, *An Environmental History of Latin America*, (Cambridge University Press, 2007), 16–18.

15 Paul Crutzen, "Geology of Mankind," *Nature* 415 (2002): 23.

16 Christophe Bonneuil and Jean-Baptiste Fressoz, *The Shock of the Anthropocene* (Verso, 2016), 180–183.

17 Eduard Suess, *The Face of the Earth*, volume 2, trans. Hertha B. C. Sollas (Clarendon, 1906), 210.

18 Gregg Mitman, Libby Robin and Sverker Sörlin offer alternative intellectual histories in *Future Remains: A Cabinet of Curiosities for the Anthropocene* (University of Chicago Press, 2017).

19 Rob Hengeveld, *Wasted World: How Our Consumption Challenges the Planet* (University of Chicago Press, 2012), 291.

20 William H. Calvin, *Global Fever: How to Treat Climate Change* (University of Chicago Press, 2008).

21 David W. Orr, *Down to the Wire: Confronting Climate Collapse* (Oxford University Press, 2009).

22 Peter Sloterdijk, "The Anthropocene: A Process-State on the Edge of Geohistory?" in *Textures of the Anthropocene: Grain, Vapor, Ray*, volume 3, ed. Katrin Klingan et al. (MIT Press, 2015), 261.

23 Johan Rockström et al., "A Safe Operating Space for Humanity," *Nature* 461 (September 24, 2009): 472–475.

24 Bill McKibben, *The End of Nature* (Doubleday, 1989), 60. The substance of his argument appeared earlier as "The End of Nature" in the September 11, 1989, issue of *The New Yorker*.

25 McKibben, *The End of Nature*, 62.

26 Elizabeth Kolbert, "The Climate of Man—III," *The New Yorker*, May 9, 2005, 52; Dipesh Chakrabarty, "The Climate of History: Four Theses," *Critical Inquiry* 35, no. 2 (2009): 197–222.

27 Donna Haraway, "Anthropocene, Capitalocene, Plantationocene, Chthulucene: Making Kin," *Environmental Humanities* 6 (2015): 161; Haraway, *Simians, Cyborgs, and Women: The Reinvention of Nature* (Routledge, 1990).

28 Chakrabarty, "The Climate of History," 218.

29 Slavoj Žižek, *Living in the End Times* (Verso, 2010), 330–336.

30 Rob Nixon, "The Anthropocene: Promises and Pitfalls of an Epochal Idea," *Edge Effects*, November 6, 2014. http://edgeeffects.net/anthropocene-promise-and-pitfalls/

31 Thomas Piketty, *Capital in the Twenty-First Century*, trans. Arthur Goldhammer (Harvard University Press, 2014).

32 For an excellent contextual discussion of the "good Anthropocene," see Lisa Sideris, "Anthropocene Convergences: A Report from the Field," *RCC Perspectives* 2016, no. 2: 89–96.

33 Eileen Crist, "On the Poverty of our Nomenclature," *Environmental Humanities* 3 (2013): 132.

34 Andreas Malm and Alf Hornborg, "The Geology of Mankind? A Critique of the Anthropocene Narrative," *Anthropocene Review* 1, no. 1 (2014): 62–69.

35 *Welcome to the Anthropocene: The Earth in Our Hands*, ed. Nina Möllers, Christian Schwägerl, and Helmuth Trischler (Rachel Carson Center/Deutsches Museum, 2014), 6.

36 Chakrabarty, "The Climate of History," 217.

37 David E. Nye and Sarah Elkind, *The Antilandscape* (Rodopi, 2014), 11–28.

38 Allan M. Winkler, *Life Under a Cloud: American Anxiety about the Atom* (Oxford University Press, 1993).

39 Gregg Mitman, Michelle Murphy, and Christopher Sellers, "A Cloud over History," *Osiris* 19 (2004): 1.

40 Ulrich Beck, *World Risk Society* (Polity, 1999); Naomi Klein, *This Changes Everything: Capitalism vs. the Climate* (Simon & Schuster, 2014).

41 The Breakthrough Institute, "An Ecomodernist Manifesto," http://www.ecomodernism.org/. For critical responses, see the special commentary section of *Environmental Humanities* 7, no. 1 (2015).

42 William B. Gail, "A New Dark Age Looms," *New York Times*, April 19, 2016.

43 Food and Agriculture Organization of the United Nations, How to Feed the World in 2050 (2009), 27.

44 http://grist.org/cities/stop-trying-to-save-the-planet-says-urban-ranger-jenny-price/

45 http://dark-mountain.net/blog/

46 Alexa Weik von Mossner, "Science Fiction and the Risks of the Anthropocene: Anticipated Transformations in Dale Pendell's The Great Bay," *Environmental Humanities* 5 (2014): 214.

47 Agnes Wooley, ""There's a Storm Coming!": Reading the Threat of Climate Change in Jeff Nichols's *Take Shelter*," *Interdisciplinary Studies in Literature and Environment* 21, no. 1 (2014): 176.

48 Lawrence Buell, *Writing for an Endangered World: Literature, Culture, and Environment in the U.S. and Beyond* (Harvard University Press, 2009), 31.

49 Svetlana Alexievich, *Voices from Chernobyl: The Oral History of a Nuclear Disaster* (Picador, 2006), 236.

50 Ibid., 91.

51 Ibid., 115.

52 Ibid., 118.

53 Ibid., 107.

54 Buell, *From Apocalypse to Way of Life*, 165.

55 *Manufactured Landscapes*, dir. Jennifer Baichwal, Foundry Films (2006).

56 "There is no document of civilization which is not at the same time a document of barbarism." Walter Benjamin, "Theses on the Philosophy of History," in *Illuminations* (Schocken, 1968), 256.

57 Glenn Albrecht et al., "Solastalgia: The Distress Caused by Environmental Change," *Australasian Psychiatry* 15 (2007): 96.

58 Martha Serpas, "Corollary," *The Dirty Side of the Storm* (Norton, 2007).

59 Lemenager, *Living Oil: Petroleum Culture in the American Century* (Oxford University Press, 2014).

60 Glenn Albrecht, "The Age of Solastalgia," *The Conversation*, August 7, 2012.

61 Erik Reece, *Lost Mountain: A Year in the Vanishing Wilderness* (Riverhead Books, 2006), 100.

Chapter 6

1 Cited in Blatt, *America's Environmental Report Card*, 222.

2 Paul Hawken, *Blessed Unrest* (Viking, 2007).

3 Paul Hawken, *The Ecology of Commerce* , revised edition (Collins Business, 2010).

4 Hawken, *Blessed Unrest*, 176–178.

5 Wendell Berry, *The Gift of Good Land: Further Essays Cultural and Agricultural* (North Point, 1983); Lawrence Buell, *Writing for an Endangered World: Literature, Culture, and Environment in the US and Beyond* (Harvard University Press, 2001), 59–81.

6 Raymond De Young and Thomas Princen, *The Localization Reader* (MIT Press, 2012).

7 Ibid., xii–xiii.

8 Ibid., xxi.

9 Joachim Radkau, *Nature and Power: A Global History of the Environment* (Cambridge University Press, 2008), 257.

10 http://gen.ecovillage.org/

11 Karen Litfin, "A Whole New Way of Life: Ecovillages and the Revitalization of Deep Community," in *The Localization Reader*, ed. Raymond De Young and Thomas Princen (MIT Press, 2012), 130–131.

12 Ibid., 136.

13 Ibid., 137.

14 Howard P. Segal, *Utopias: A Brief History from Ancient Writings to Virtual Communities* (Wiley, 2012), 24–28.

15 Andrew Simms, "Sønderborg: The Little-Known Danish Town with a Zero Carbon Master Plan," *The Guardian,* October 22, 2015. http://vaekstraad.sonderborg.

dk/ http://www.projectzero.dk/da-DK/Artikler/2014/December/L%C3%A6r-om-gr%C3
%B8n-omstilling-i-S%C3%B8nderborg.aspx

16 http://www.energibyenskive.dk/da/om-energibyen-skive/borgmesterpagten/
government-gazelle/

17 David Hess, "Global Problems, Localist Solutions," in *The Localization Reader*, ed.
De Young and Princen, 272.

18 Baris Gencer Baykan, "From Limits of Growth to Degrowth within French Green
Politics," *Environmental Politics* 16, no. 3 (2017): 514–515.

19 "A History of Degrowth," http://degrowth.de/en/a-history-of-degrowth.

20 http://alternation.info/can-decroix-a-research-degrowth-community-project/

21 http://www.newcitiesfoundation.org/making-cities-self-sufficient-food-production/

22 Peter Dauvergne, *The Shadows of Consumption: Consequences for the Global Environment* (MIT Press, 2008), 8.

23 Ibid., 14.

24 Posted November 2, 2015, at degrowth.org/.

25 Joseph Stiglitz, *The Roaring '90s* (Norton, 2003), 124.

26 David E. Nye, *America's Assembly Line* (MIT Press, 2013), 221–239.

27 Paul Hawken and Amory Lovins, *Natural Capitalism* (Earthscan, 1999).

28 William McDonough, and Michael Braungart, *Cradle to Cradle* (North Point,
2002).

29 William McDonough and Michael Braungart, *The Upcycle: Beyond Sustainability—
Designing for Abundance* (North Point, 2013).

30 See James Fleming, "Skyscapes and Anti-skyscapes: Making the Invisible Visible,"
in *The Antilandscape*, ed. David E. Nye and Sarah Elkind (Rodopi, 2014), 40–41.

31 https://living-future.org/case-study/csl

32 Cited in Stephanie LeMenager, *Living Oil* (Oxford University Press, 2014), 12.

33 Martin Heidegger, *The Question Concerning Technology and Other Essays,* trans.
William Lovitt (Harper, 1982), 21.

34 LeMenager, *Living Oil*, 13.

35 Ben Tufnell, *Land Art* (Tate, 2006), 16.

36 Alan Sekula, *Fish Story* (Witte de With and Richter, 1995).

37 John Lanchester, *I.O.U.: Why Everyone Owes Everyone and No One Can Pay* (Simon and Schuster), 67.

38 Philip Lawn, "A Stock Take of Green National Accounting Initiatives," *Social Indicators Research* 80, no. 2 (2007): 427–460.

39 Wouter van Dieren, ed., *Taking Nature into Account A Report to the Club of Rome Toward a Sustainable National Income* (Springer, 1995), 236.

40 Walter Radermacher, "Indicators, Green Accounting and Environment Statistics: Information Requirements for Sustainable Development," *International Statistical Review* 67, no. 3 (1999): 339–354.

41 Lawn, "A Stock Take of Green National Accounting Initiatives," 447–448.

42 Bill McKibben, *Deep Economy: The Wealth of Communities and the Durable Future* (Times Books, 2007).

43 Naomi Klein, *This Changes Everything: Capitalism vs. the Climate* (Simon & Schuster, 2014).

44 Jeffrey Bolster, *The Mortal Sea: Fishing the Atlantic in the Age of Sail* (Harvard University Press, 2012), 278.

45 *Wall Street Journal,* May 2, 2013.

46 http://www.usgbc.org/articles/green-building-facts

47 http://www.un.org/sustainabledevelopment/climate-change-2/

Chapter 7

1 Adrian Parr, *Hijacking Sustainability* (MIT Press, 2009).

2 Rick Dolphijn and Iris van der Tuin, *New Materialism: Interviews and Cartographies* (Open Humanities Press, 2012), 13.

3 Timothy Morton, *The Ecological Thought* (Harvard University Press, 2010), 28.

4 Tim Lecain, "Against the Anthropocene: A Neo-Materialist Perspective," *International Journal for History, Culture and Modernity* 3, no. 1 (2015): 4.

5 Kate Rigby, "Spirits That Matter: Pathways towards a Rematerialization of Religion and Spirituality," in *Material Ecocriticism*, ed. Serenella Iovino and Serpil Oppermann (Indiana University Press, 2014), 283.

6 Jason W. Moore, *Capitalism in the Web of Life* (Verso, 2015), 75.

7 Ibid., 172.

8 Ibid., 305.

9 Ibid., 304.

10 Jane Bennett, *Vibrant Matter: A Political Ecology of Things* (Duke University Press, 2010), xix.

11 Ibid., 29.

12 Ibid., 39.

13 Ibid., 108.

14 Bruno Latour, *Politics of Nature: How to Bring the Sciences into Democracy*, trans. Catherine Porter (Harvard University Press, 2009), 60.

15 Bennett, *Vibrant Matter*, 122.

16 Janell Watson, "Eco-Sensibilities: An Interview with Jane Bennett," *Minnesota Review* 81 (2013): 147–148.

17 Heather Houser, *Ecosickness in Contemporary U.S. Fiction: Environment and Affect* (Columbia University Press, 2014), 8.

18 Ibid., 5.

19 See Serenella Iovino and Serpil Oppermann, *Material Ecocriticism* (Indiana University Press, 2014) and the "Material Ecociticism" special issue of *Interdisciplinary Studies of Literature and Environment* edited by Dana Phillips and Heather Sullivan (19, no. 2 [2012]).

20 Heather Sullivan, "Dirt Theory and Material Ecocriticism," *Interdisciplinary Studies of Literature and Environment* 19, no. 3 (2012): 515; Stacy Alaimo, "Violet Black," in *Prismatic Ecologies: Ecotheory Beyond Green*, ed. Jeffrey Cohen (University of Minnesota Press, 2013), 233.

21 Susan Naramore Maher, *Deep Map Country: Literary Cartography of the Great Plains* (University of Nebraska Press, 2014).

22 Jim Cheney, "Truth, Knowledge, and the Wild World," *Ethics and Environment* 10, no. 2 (2005): 108.

23 Ibid., 120.

24 Rinda West, *Out of the Shadow: Ecopsychology, Story, and Encounters with the Land* (University of Virginia Press, 2007), 2.

25 Ibid., 45.

26 Ibid., 2.

27 Al Gedicks, *Resource Rebels: Native Challenges to Mining and Oil Corporations* (South End, 2001), 41.

28 Walter Rodney, *How Europe Underdeveloped Africa* (Bogle-L'Ouverture, 1972).

29 On the concept of "ecosystem people," see Ramachandra Guha, *How Much Should a Person Consume? Environmentalism in India and the United States* (University of California Press, 2006), 233–236.

30 Clapperton Chakanetsa Mavhunga, *Transient Workspaces: Technologies of Everyday Innovation in Zimbabwe* (MIT Press, 2014), 7.

31 The term "environmentalism of the poor" owes much to Guha's early work on the Chipko or tree-hugging movement in northern India. Michael Watts and Richard Peet developed a parallel concept, "liberation ecology"; see *Liberation Ecologies: Environment, Development, Social Movements*, ed. Watts and Peet (Routledge, 2004).

32 Joan Martinez Alier, *The Environmentalism of the Poor: A Study of Ecological Conflicts and Valuation* (Elgar, 2002), 12.

33 Beth Rose Middleton, *Trust in the Land: New Directions in Tribal Conservation* (University of Arizona Press, 2011), 35.

34 Arn Keeling and John Sandlos, "Introduction: Critical perspectives on extractive industries in Northern Canada," *The Extractive Industries and Society* 3 (2016): 266. On the legacies of mining in northern Canada for First Nations, see the full special issue of *The Extractive Industries and Society*.

35 Winona LaDuke, *All Our Relations: Native Struggles for Land and Life* (South End, 1999), 98.

36 Middleton, *Trust in the Land*, 46–56.

37 Joni Adamson, *American Indian Literature, Environmental Justice, and Ecocriticism: The Middle Place* (University of Arizona Press, 2001); *New Perspectives on Environmental Justice: Gender, Sexuality, and Activism*, ed. Rachel Stein (Rutgers University Press, 2004), 1.

38 Naomi Klein, *This Changes Everything: Capitalism Versus the Climate* (Simon and Schuster, 2014), 443.

39 Charlotte Coté, *Spirits of Our Whaling Ancestors: Revitalizing Makah and Nuu-chah-nulth Traditions* (University of Washington Press, 2010), 6.

40 Ibid., 199.

41 William Bartram, *Travels and Other Writings*, ed. Thomas Slaughter (Library of America, 1996).

42 C. Margaret Scarry, "Patterns of Wild Plant Utilization," in *People and Plants in Ancient Eastern North America*, ed. Paul E. Minnis (Smithsonian Books, 2003), 76, 50–104.

43 J. Baird Callicott, *Earth's Insights* (University of California Press, 1994), 124.

44 Ibid., 111.

45 Shepard Krech III, *The Ecological Indian: Myth and History* (Norton, 1999), 28.

46 Though it is best to begin by reading Krech's monograph, his response to the controversy is also instructive. See Shepard Krech III, "Reflections on Conservation, Sustainability, and Environmentalism in Indigenous North America," in *American Anthropologist* 107, no. 1 (2005): 78–86.

47 Kimberly TallBear, "Shepard Krech's *The Ecological Indian*: One Indian's Perspective" *International Institute for Indigenous Resource Management Publications* http://www.iiirm.org/publications/Book%20Reviews/Reviews/Krech001.pdf (September 2000), 4.

48 Emmanuel Lévinas, *Otherwise Than Being, or Beyond Essence*, trans. Alphonso Lingis (Duquesne University Press, 1998).

49 Walt Whitman, "Song of Myself," in *Leaves of Grass*, ed. Sculley Bradley and Harold W. Blodgett (Norton, 1973), stanza 32 in the revised 1882 version of the poem, although the text is substantially the same in the 1855 version, published four years before Darwin's *Origin of Species* appeared in print.

50 Andrew Linzey, ed., *The Global Guide to Animal Protection* (University of Illinois Press, 2013), 43–44.

51 Deborah Bird Rose, "Val Plumwood's Philosophical Animism: Attentive Interactions in the Sentient World," *Environmental Humanities* 3 (2013): 100.

52 Tom Regan, *The Case for Animal Rights*, updated edition (University of California Press, 2004), 243.

53 Warwick Fox, *A Theory of General Ethics: Human Relationships, Nature, and the Built Environment* (MIT Press, 2006), 22–32.

54 Peter Singer, *Animal Liberation* (Harper Perennial, 2009). See also Peter Singer, ed., *In Defense of Animals* (Blackwell, 1985).

55 Paul Warner, *Living Through the End of Nature* (MIT Press, 2010), 118.

56 Lisa Kemmerer, ed., *Sister Species: Women Animals, and Social Justice* (University of Illinois Press, 2011).

57 Ibid., 184.

58 Stacy Alaimo, *Bodily Natures: Science, Environment, and the Material Self* (Indiana University Press, 2010).

59 Evelyn Fox Keller, *Reflections on Gender and Science* (Yale University Press, 1985), 117.

60 Timothy Morton, "Guest Column: Queer Ecology," *PMLA* 125, no. 2 (2010): 274.

61 Greta Gaard, "Toward a Queer Ecofeminism," *Hypatia* 12, no. 1 (1997): 114–137.

62 Catriona Sandilands, "Sexual Politics and Environmental Justice," in *New Perspectives on Environmental Justice: Gender, Sexuality and Activism*, ed. Rachel Stein (Rutgers University Press, 2004), 109.

63 On the commodification of queer identities and "ecophobia," see Simon Estok, "Theorizing in a Space of Ambivalent Openness: Ecocriticism and Ecophobia," *ISLE* 16, no. 2 (2009): 203–225.

64 Morton, "Guest Column: Queer Ecology," 276.

65 Lee Edelman, *No Future: Queer Theory and the Death Drive* (Duke University Press, 2004).

66 Robert Azarello, *Queer Environmentality: Ecology, Evolution, and Sexuality in American Literature* (Ashgate, 2012), 136.

67 For a complementary argument for a care ethic rooted in ecological citizenship rather than constraining, heteronormative roles assigned to women-as-mothers, see Sherilyn MacGregor, "From Care to Citizenship: Calling Ecofeminism Back to Politics," *Ethics and the Environment* 9, no. 1 (2004): 56–84.

68 Nicole Seymour, *Strange Natures: Futurity, Empathy, and the Queer Ecological Imagination* (Illinois University Press, 2013), 27.

69 Matthew Gandy, "Queer Ecology: Nature, Sexuality, and Heterotopic Alliances," *Environment and Planning D* 30 (2012): 727.

70 Ibid., 738.

71 David Abram, "The Commonwealth of Breath," in *All Our Relations: 18th Biennale of Sydney 2012*, ed. Catherine de Zegher and Gerald McMaster (Biennale of Sydney Ltd., 2012), 341.

Chapter 8

1 Lewis Mumford, *The Myth of the Machine* (Harcourt Brace Jovanovich, 1979), 264.

2 http://environment.princeton.edu/grandchallenges/

3 David E. Boyd, *The Optimistic Environmentalist: Progressing Toward a Greener Future* (ECW, 2015), 91–102.

4 "Shell in the Arctic," a spoof produced by Greenpeace and the Yes Men, can still be viewed at the Internet Archive in its 2012 version: https://web.archive.org/web/20120612154007/http://arcticready.com/

5 Nanang Indra Kurniawan and Ståle Angen Rye, "Online Environmental Activism and Internet Use in the Indonesian Environmental Movement," *Information Development* 30, no. 3 (2014): 200–212.

6 Jonathan Sullivan and Lei Xie, "Environmental Activism, Social Networks and the Internet," *China Quarterly*, June 2009.

7 Rex Weyler, "Nature's Apprentice: A Meta-Narrative for Aging Empires," *Manoa* 25, no. 1: 187–196.

8 Sverker Sörlin, "Environmental Humanities: Why Should Biologists Interested in the Environment Take the Humanities Seriously?" *BioScience* 62, no. 9 (2012): 788–789.

9 Eduardo Kohn, *How Forests Think: Toward an Anthropology Beyond the Human* (University of California Press, 2013); Val Plumwood, "Surviving a Crocodile Attack," *Utne Reader* (July-August 2000), reproduced from *The Ultimate Journey* (1999).

10 Hannes Bergthaller, Rob Emmett, Adeline Johns-Putra, Agnes Kneitz, Susanna Lidström, Shane McCorristine, Isabel Pérez Ramos, Dana Phillips, Kate Rigby, and Libby Robin, "Mapping Common Ground: Ecocriticism, Environmental History, and the Environmental Humanities," *Environmental Humanities* 5 (2014): 262.

11 Ibid., 272.

12 Barbara Allen, "Narrating the Toxic Landscape in 'Cancer Alley' Louisiana," in *Technologies of Landscape: From Reaping to Recycling*, ed. David E. Nye (University of Massachusetts Press, 2000).

13 Barbara Allen, *Uneasy Alchemy* (MIT Press, 2003), 117.

14 Steve Lerner, *Sacrifice Zones* (MIT Press, 2010), 113.

15 Ibid., 313.

16 David Naguib Pellow, *Resisting Global Toxics: Transnational Movements for Environmental Justice* (MIT Press, 2007), 226.

17 http://therightsofnature.org/rights-of-nature-tribunal/; see also the online collection "Rights of Nature Recognition: Law and Ethics in Dialogue" curated by María Valeria Berros and Anna Leah Tabios, http://www.environmentandsociety.org/arcadia/collection/rights-nature-recognition

18 Chris Hedges and Joe Sacco, *Days of Destruction, Days of Revolt* (Perseus Books, 2012), 226.

19 Derek Wall, *The Commons in History: Culture, Conflict and Ecology* (MIT Press, 2014), 132–133.

Bibliography

Abbate, Janet. *Inventing the Internet*. MIT Press, 1999.

Abram, David. "The Commonwealth of Breath." In *All Our Relations: 18th Biennale of Sydney 2012*, ed. Catherine de Zegher and Gerald McMaster. Biennale of Sydney Ltd., 2012.

Adamson, Joni. *American Indian Literature, Environmental Justice, and Ecocriticism: The Middle Place*. University of Arizona Press, 2001.

Adamson, Joni, and Slovic Scott. "The Shoulders We Stand On: An Introduction to Ethnicity and Ecocriticism." *MELUS* 34, no. 2 (2009): 5–24.

Adas, Michael. *Machines as the Measure of Men: Science, Technology, and Ideologies of Western Dominance*. Cornell University Press, 1989.

Agnoletti, Mauro. "Rural Landscape, Nature Conservation and Culture: Some Notes on Research Trends and Management Approaches from a (Southern) European Perspective." *Landscape and Urban Planning* 126 (2014): 66–73.

Alaimo, Stacy. *Bodily Natures: Science, Environment, and the Material Self*. Indiana University Press, 2010.

Alaimo, Stacy. "Violet-Black: Ecologies of the Abyssal Zone." In *Prismatic Ecologies: Ecotheory Beyond Green*, ed. Jeffrey Cohen. University of Minnesota Press, 2013.

Albrecht, Glenn. "The Age of Solastalgia." *The Conversation*, August 7, 2012. https://theconversation.com/the-age-of-solastalgia-8337

Albrecht, Glenn, et al. "Solastalgia: The Distress Caused by Environmental Change." *Australasian Psychiatry* 15 (2007): 95–98.

Alexievich, Svetlana. *Voices from Chernobyl: The Oral History of a Nuclear Disaster*, trans. K. Gessen. Picador, 2006.

Alier, Joan Martinez. *The Environmentalism of the Poor: A Study of Ecological Conflicts and Valuation*. Edward Elgar, 2002.

Allen, Barbara. "Narrating the Toxic Landscape in 'Cancer Alley,' Louisiana." In *Technologies of Landscape: From Reaping to Recycling*, edited by David E. Nye, 187–203. University of Massachusetts Press, 2000.

Allen, Barbara. *Uneasy Alchemy: Citizens and Experts in Louisiana's Chemical Corridor Disputes*. MIT Press, 2003.

Alvear, Cecilia. "Are We Loving the Galápagos to Death?" *Harvard Review of Latin America* 8, no. 3 (2009): 22–23.

Azarello, Robert. *Queer Environmentality: Ecology, Evolution, and Sexuality in American Literature*. Ashgate, 2012.

Barrow, Mark. *Nature's Ghosts: Confronting Extinction from the Age of Jefferson to the Age of Ecology*. University of Chicago Press, 2009.

Bartram, William. *Travels and Other Writings*. Edited by Thomas Slaughter. Library of America, 1996.

Baykan, Baris Gencer. "From Limits of Growth to Degrowth within French Green Politics." *Environmental Politics* 16 (2017): 3.

Beck, Ulrich. *World Risk Society*. Polity, 1999.

Belasco, Warren J. *Appetite for Change: How the Counterculture Took on the Food Industry*. Cornell University Press, 1993.

Benjamin, Walter. *Illuminations*. Schocken, 1968.

Bennett, Jane. *Vibrant Matter: A Political Ecology of Things*. Duke University Press, 2010.

Bergthaller, Hannes, Rob Emmett, Adeline Johns-Putra, Agnes Kneitz, Susanna Lidström, Shane McCorristine, Isabel Pérez Ramos, Dana Phillips, Kate Rigby, and Libby Robin. "Mapping Common Ground: Ecocriticism, Environmental History, and the Environmental Humanities." *Environmental Humanities* 5 (2014): 261–276.

Berlant, Lauren. *Cruel Optimism*. Duke University Press, 2011.

Berry, Wendell. *The Gift of Good Land*. North Point, 1982.

Berry, Wendell. *The Unsettling of America*. Sierra Club, 1977.

Blatt, Harvey. *America's Environmental Report Card*. MIT Press, 2005.

Bolster, Jeffrey. *The Mortal Sea: Fishing the Atlantic in the Age of Sail*. Harvard University Press, 2012.

Bonneuil, Christophe, and Jean-Baptiste Fressoz. *The Shock of the Anthropocene*. Verso, 2016.

Boyd, David E. *The Optimistic Environmentalist: Progressing Towards a Greener Future*. ECW, 2015.

Buckley, Michael. *Meltdown in Tibet: China's Reckless Destruction of the Ecosystems from the Highlands of Tibet to the Deltas of Asia.* Palgrave Macmillan, 2014.

Buell, Frederick. *From Apocalypse to Way of Life: Environmental Crisis in the American Century.* Routledge, 2003.

Buell, Lawrence. *Writing for an Endangered World: Literature, Culture, and Environment in the U.S. and Beyond.* Harvard University Press, 2009.

Bullard, R., ed. *The Quest for Environmental Justice.* Counterpoint, 2005.

Buzard, James. *The Beaten Track: European Tourism, Literature, and the Ways to "Culture," 1800–1918.* Clarendon, 1993.

Callicott, J. Baird. *Earth's Insights.* University of California Press, 1994.

Calvin, William H. *Global Fever: How to Treat Climate Change.* University of Chicago Press, 2008.

Carney, Judith A., and Richard Nicholas Rosomoff. *In the Shadow of Slavery: Africa's Botanical Legacy in the Atlantic World.* University of California Press, 2010.

Carrigan, Anthony. "'Out of This Great Tragedy Will Come a World Class Tourism Destination': Disaster, Ecology, and Post-Tsunami Tourism Development in Sri Lanka." In *Postcolonial Ecologies: Literatures of the Environment*, ed. Elizabeth DeLoughrey and George B. Handley, 273–290. Oxford University Press, 2011.

Carruthers, Jane. *The Kruger National Park: A Social and Political History.* University of Natal Press, 1995.

Carson, Rachel. *The Edge of the Sea.* Houghton Mifflin, 1955.

Carson, Rachel. *Silent Spring.* Houghton Mifflin, 1962.

Chakrabarty, Dipesh. "The Climate of History: Four Theses." *Critical Inquiry* 35, no. 2 (2009): 197–222.

Cheney, Jim. "Truth, Knowledge, and the Wild World." *Ethics and the Environment* 10, 2005: 2.

Cohen, Lisbeth. "Encountering Mass Culture at the Grassroots: The Experience of Chicago Workers in the 1920s." *American Quarterly* 41, no. 1 (1989): 6–33.

Cohen, Lisbeth. *A Consumer's Republic.* Knopf, 2003.

Conte, Christopher. *Highland Sanctuary: Environment and History in Tanzania's Usambara Mountains.* Ohio University Press, 2004.

Coté, Charlotte. *Spirits of Our Whaling Ancestors: Revitalizing Makah and Nuu-chah-nulth Traditions.* University of Washington Press, 2010.

Cowan, Nancy. *Peregine Spring.* Rowman & Littlefield, 2016.

Cowan, Ruth Schwartz. *More Work for Mother*. Basic Books, 1983.

Crist, Eileen. "On the Poverty of Our Nomenclature." *Environmental Humanities* 3 (2013): 129–147.

Cronon, William. "The Trouble with Wilderness: Or, Getting Back to the Wrong Nature." In *Uncommon Ground: Rethinking the Human Place in Nature*. Norton, 1995, 69–90.

Crosby, Alfred W. *Ecological Imperialism*. Cambridge University Press, 2015.

Crutzen, Paul. "Geology of Mankind." *Nature* 415 (2002): 23.

Crutzen, Paul, and Eugene Stoermer. "The 'Anthropocene.'" *IGBP Newsletter*, no. 41, 2000: 17–18.

Dauvergne, Peter. *The Shadows of Consumption: Consequences for the Global Environment*. MIT Press, 2008.

DeLillo, Don. *White Noise*. Penguin Classsics, 2009.

DeLoughrey, Elizabeth, and George B. Handley, "Introduction: Toward an Aesthetics of the Earth," In *Postcolonial Ecologies: Literatures of the Environment*, ed. Elizabeth DeLoughrey and George B. Handley. Oxford University Press, 2011.

DeLoughrey, E., and G. B. Handley, eds. *Postcolonial Ecologies: Literatures of the Environment*. Oxford University Press, 2011.

Deutch, John M. *The Crisis in Energy Policy*. Harvard University Press, 2011.

De Young, Raymond, and Thomas Princen. *The Localization Reader*. MIT Press, 2012.

Dolphijn, Rick, and Iris van der Tuin. *New Materialism: Interviews and Cartographies*. Open Humanities Press, 2012.

Doyle, Timothy. *Environmental Movements*. Rutgers University Press, 2005.

Edelman, Lee. *No Future: Queer Theory and the Death Drive*. Duke University Press, 2004.

Eisenberg, Christina. *The Carnivore Way: Coexisting with and Conserving North America's Predators*. Island, 2014.

Emmett, Robert S. *Cultivating Environmental Justice: A Literary History of U.S. Garden Writing*. University of Massachusetts Press, 2016.

Estok, Simon. "Theorizing in a Space of Ambivalent Openness: Ecocriticism and Ecophobia." *ISLE* 16, no. 2 (2009): 203–225.

Fahn, James David. *A Land on Fire: The Environmental Consequences of the Southeast Asian Boom*. Westview, 2003.

Fleming, James Rodger. *Fixing the Sky: The Checkered History of Weather and Climate Control*. Columbia University Press, 2010.

Fleming, James Rodger. "Skyscapes and Anti-skyscapes: Making the Invisible Visible." In *The Antilandscape*, ed. David E. Nye and Sarah Elkind. Rodopi, 2014.

Foucault, Michel. *The Birth of Biopolitics*, ed. Michel Senellart. Palgrave Macmillan, 2008.

Fox, Warwick. *A Theory of General Ethics: Human Relationships, Nature, and the Built Environment*. MIT Press, 2006.

Gaard, Greta. "Toward a Queer Ecofeminism." *Hypatia* 12, no. 1 (1997): 114–137.

Gadgil, Madhav, and Ramachandra Guha. *This Fissured Land: An Environmental History of India*. University of California Press, 1993.

Gadgil, Madhav, and Ramachandra Guha. *The Use and Abuse of Nature*. Oxford University Press, 2000.

Gandy, Matthew. "Queer Ecology: Nature, Sexuality, and Heterotopic Alliances." *Environment and Planning D* 30 (2012): 727–747.

Ganopolski, Andrey, Ricarda Winkelman, and Hans Joachim Schellnhuber. "Critical Insolation–CO_2 Relation for Diagnosing Past and Future Glacial Inception." *Nature* 529 (January 14, 2016): 200–203.

Gardiner, Stephen M. "Some Early Ethics of Geoengineering the Climate: A Commentary on the Values of the Royal Society Report." *Environmental Values* 20 (2011): 163–188.

Gaynor, Andrea. *Harvest of the Suburbs: An Environmental History of Growing Food in Australian Cities*. University of Western Australia Press, 2006.

Gedicks, Al. *Resource Rebels: Native Challenges to Mining and Oil Corporations*. South End, 2001.

Ghosh, Amitav. *The Hungry Tide*. Houghton Mifflin Harcourt, 2005.

Goffman, Erving. *The Presentation of Self in Everyday Life*. Anchor Books, 1959.

Goldberg, Jeffrey. "Were There Dinosaurs on Noah's Ark?" *The Atlantic*, October 2014.

Griffiths, Tom. "The Humanities and an Environmentally Sustainable Australia." *Australian Humanities Review*, December 2007. Australianhumanitiesreview.org

Griffiths, Tom, and Libby Robin, eds. *Ecology and Empire: Environmental History of Settler Societies*. University of Washington Press, 1997.

Grimm, Ruediger H. *Nietzsche's Theory of Knowledge*. Walter de Gruyter, 1977.

Grove, Richard. *Green Imperialism: Colonial Expansion, Tropical Island Edens and the Origins of Environmentalism, 1600–1860.* Cambridge University Press, 1996.

Grove, Richard, and Vinita Damodaran. 2009. "Imperialism, Intellectual Networks, and Environmental Change: Unearthing the Origins and Evolution of Global Environmental History." In *Nature's End: History and Environment,* ed. Sverker Sörlin and Paul Warde, 23–49. Palgrave Macmillan, 2009.

Guha, Ramachanda. *How Much Should a Person Consume? Environmentalism in India and the United States.* University of California Press, 2006.

Hamilton, Clive. *Earthmasters: The Dawn of the Age of Climate Engineering.* Yale University Press, 2013.

Haraway, Donna. "A Manifesto for Cyborgs." In *The Haraway Reader.* Routledge, 2004.

Haraway, Donna. *Simians, Cyborgs, and Women: The Reinvention of Nature.* Routledge, 1990.

Haraway, Donna. *When Species Meet.* University of Minnesota Press, 2008.

Haraway, Donna. "Anthropocene, Capitalocene, Plantationocene, Chthulucene: Making Kin." *Environmental Humanities* 6 (2015): 161.

Hawken, Paul. *Blessed Unrest.* Viking, 2007.

Hawken, Paul. *The Ecology of Commerce.* Collins Business, 2010.

Hawken, Paul, and Amory Lovins. *Natural Capitalism.* Earthscan, 1999.

Hayles, Katherine. *How We Became Posthuman: Virtual Bodies in Cybernetics, Literature, and Informatics.* University of Chicago Press, 1999.

Hedges, Chris, and Joe Sacco. *Days of Destruction, Days of Revolt.* Perseus Books, 2012.

Heise, Ursula. "The Hitchhiker's Guide to Ecocriticism." *PMLA* 121, no. 2 (2006): 503–516.

Heise, Ursula. *Sense of Place, Sense of Planet: The Environmental Imagination of the Global.* Oxford University Press, 2008.

Hengeveld, Rob. *Wasted World: How Our Consumption Challenges the Planet.* University of Chicago Press, 2012.

Hess, David. "Global Problems, Localist Solutions." In *The Localization Reader,* ed. Raymond De Young and Thomas Princen. MIT Press, 2012.

Holling, C. S. "Resilience and Stability of Ecological Systems." *Annual Review of Ecology and Systematics* 4 (1973): 1–23.

Holm, Poul, et al. "Collaboration between the Natural, Social and Human Sciences in Global Change Research." *Environmental Science & Policy* 28 (2013): 25–35.

Hou, Jeffrey. *Insurgent Public Space: Guerrilla Urbanism and the Remaking of Contemporary Cities.* Routledge, 2010.

Houser, Heather. *Ecosickness in Contemporary U.S. Fiction: Environment and Affect.* Columbia University Press, 2014.

Hulme, M., ed. *Climates and Cultures.* SAGE, 2015.

Huxley, Aldous. *Brave New World.* Chatto and Windus, 1932.

Iovino, Serenella, and Serpil Oppermann. *Material Ecocriticism.* Indiana University Press, 2014.

Jacoby, Karl. *Crimes Against Nature: Squatters, Poachers, Thieves, and the Hidden History of American Conservation.* University of California Press, 2014.

Jensen, Derek. *What We Leave Behind.* Penguin, 2009.

Jørgensen, Dolly. "Migrant Muskoxen and the Naturalization of National Identity in Scandinavia." In *The Historical Animal,* ed. S. Nance. Syracuse University Press, 2015.

Kalevi, Kull. Personal communication during Animals in Transdisciplinary Environmental History ESEH Summer School, Haapsalu, Estonia, May 2015.

Keeling, Arn, and John Sandlos. "Introduction: Critical Perspectives on Extractive Industries in Northern Canada." *Extractive Industries and Society* 3 (2016): 265–268.

Keller, Evelyn Fox. *Reflections on Gender and Science.* Yale University Press, 1985.

Kelley, Klara Bonsack, and Harris Francis. *Navajo Sacred Places.* Indiana University Press, 1994.

Kelly, Kevin. *What Technology Wants.* Viking, 2010.

Kemmerer, L., ed. *Sister Species: Women, Animals, and Social Justice.* University of Illinois Press, 2011.

Kendal, Jeremy, Jamshid J. Tehrani, and John Odling-Smee. "Human Niche Construction in Interdisciplinary Focus." *Philosophical Transactions of the Royal Society of London. Series B, Biological Sciences* 366 (2011): 785–792.

Kirksey, Eben. *Emergent Ecologies.* Duke University Press, 2015.

Kirksey, Eben, and Stefan Helmreich. "The Emergence of Multispecies Ethnography: A Special Guest-Edited Issue of Cultural Anthropology." *Cultural Anthropology* 25, no. 4 (2010): 545–576.

Klein, Naomi. *This Changes Everything: Capitalism vs. the Climate.* Simon and Schuster, 2014.

Kohn, Eduardo. *How Forests Think: Toward an Anthropology Beyond the Human.* University of California Press, 2013.

Kolbert, Elizabeth. "The Climate of Man—III." *The New Yorker,* May 9, 2005.

Kolbert, Elizabeth. "Recall of the Wild: The Quest to Engineer a World Before Humans." *The New Yorker,* December 24 and 31, 2012: 24–31.

Kolbert, Elizabeth. *The Sixth Extinction.* Bloomsbury Books, 2014.

Krech, Shepard, III. *The Ecological Indian: Myth and History.* Norton, 1999.

Krech, Shepard, III. "Reflections on Conservation, Sustainability, and Environmentalism in Indigenous North America." *American Anthropologist* 107, no. 1 (2005): 78–86.

Kurniawan, Nanang Indra, and Ståle Angen Rye. "Online Environmental Activism and Internet Use in the Indonesian Environmental Movement." *Information Development* 30, no. 3 (2014): 200–212.

Kurzweil, Ray. *The Singularity Is Near.* Viking, 2005.

Lachmund, Jens. *Greening Berlin: The Co-production of Science, Politics, and Urban Nature.* MIT Press, 2013.

LaDuke, Winona. *All Our Relations: Native Struggles for Land and Life.* South End, 1999.

Laitner, John A., and Karen Ehrhardt-Martinez. "Information and Communication Technologies: The Power of Productivity." Report E081 from Washington: American Council for an Energy Efficient Economy, February 2008.

Larson, Christina. "China's Grand Plans for Eco Cities Now Lie Abandoned." *Yale Environment 360,* April 6, 2009.

Latour, Bruno. *Politics of Nature: How to Bring the Sciences into Democracy.* Harvard University Press, 2009.

Lawn, Philip. "A Stock-Take of Green National Accounting Initiatives." *Social Indicators Research* 80, no. 2 (2007): 427–460.

Lecain, Tim. "Against the Anthropocene: A Neo-Materialist Perspective." *International Journal for History, Culture and Modernity* 3, no. 1 (2015): 1–28.

Lemenager, Stephanie. *Living Oil: Petroleum Culture in the American Century.* Oxford University Press, 2014.

Leopold, Aldo. *A Sand County Almanac With Essays on Conservation from Round River.* Ballantine, 1970.

Lerner, Steve. *Sacrifice Zones.* MIT Press, 2010.

Lévinas, Emmanuel. *Otherwise Than Being, or Beyond Essence,* trans. A. Lingis. Duquesne University Press, 1998.

Linzey, A., ed. *The Global Guide to Animal Protection.* University of Illinois Press, 2013.

Liotta, P. H., and W. Allan Shearer. *Gaia's Revenge: Climate Change and Humanity's Loss*. Praeger, 2007.

Litfin, Karen. "A Whole New Way of Life: Ecovillages and the Revitalization of Deep Community." In *The Localization Reader*, ed. Raymond De Young and Thomas Princen. MIT Press, 2012.

Lopez, Barry. *Crossing Open Ground*. Vintage Books, 1989.

Lousley, Cheryl. "E. O. Wilson's Biodiversity, Commodity Culture, and Sentimental Globalism." *RCC Perspectives*, no. 9, 2012: 11–16.

MacCannell, Dean. *The Tourist*. Schocken, 1989.

MacGregor, Sherilyn. "From Care to Citizenship: Calling Ecofeminism Back to Politics." *Ethics and the Environment* 9, no. 1 (2004): 56–84.

MacLellan, Matthew. "The Tragedy of Limitless Growth: Re-interpreting the Tragedy of the Commons for a Century of Climate Change." *Environmental Humanities* 7 (2015): 41–58.

Maher, Susan Naramore. *Deep Map Country: Literary Cartography of the Great Plains*. University of Nebraska Press, 2014.

Malm, Andreas, and Alf Hornborg. "The Geology of Mankind? A Critique of the Anthropocene Narrative." *Anthropocene Review* 1, no. 1 (2014): 62–69.

Manufactured Landscapes, dir. Jennifer Baichwal. Foundry Films (2006) 90 min.

Margulis, Lynn. *Symbiotic Planet: A New Look at Evolution*. Basic Books, 1998.

Marris, Emma. *The Rambunctious Garden: Saving Nature in a Post-wild World*. Bloomsbury, 2011.

Marsh, George Perkins. *Man and Nature*. University of Washington Press, 2003.

Matthiessen, Peter. *Wildlife in America*. Viking, 1959.

Mavhunga, Clapperton Chakanetsa. *Transient Workspaces: Technologies of Everyday Innovation in Zimbabwe*. MIT Press, 2014.

Mayr, Ernst. *This Is Biology: The Science of the Living World*. Harvard University Press, 1998.

McCalman, Iain. *The Reef: A Passionate History*. Scientific American / Farrar, Straus and Giroux, 2013.

McDonough, William, and Michael Braungart. *Cradle to Cradle*. North Point, 2002.

McDonough, William, and Michael Braungart. *The Upcycle: Beyond Sustainability—Designing for Abundance*. North Point, 2013.

McKibben, Bill. *The End of Nature*. Doubleday, 1989.

McKibben, Bill. *Hope, Human and Wild*. Milkweed, 2007.

McKibben, Bill. *Deep Economy: The Wealth of Communities and the Durable Future*. Times Books, 2007.

McKibben, Bill, ed. *American Earth: Environmental Writing Since Thoreau*. Library of America, 2008.

McLaren, Duncan, and Julian Agyeman. *Sharing Cities: A Case for Truly Smart and Sustainable Cities*. MIT Press, 2015.

McNeill, John. *Something New Under the Sun: An Environmental History of the Twentieth Century World*. Norton, 2000.

Melosi, Martin. "Equity, Eco-Racism and Environmental History." *Environmental History Review* 19, no. 3 (1995): 1–16.

Merchant, Carolyn. "Reinventing Eden." In *Uncommon Ground*, ed. William Cronon, 132–159. Norton, 1996.

Merton, Thomas. *Conjectures of a Guilty Bystander*. Doubleday, 1966.

Micklin, Philip P. "Desiccation of the Aral Sea: A Water Management Disaster in the Soviet Union." *Science* 241 (September 2, 1988): 1170–1176.

Middleton, Beth Rose. *Trust in the Land: New Directions in Tribal Conservation*. University of Arizona Press, 2011.

Miller, Shawn William. *An Environmental History of Latin America*. Cambridge University Press, 2007.

Mitchell, Timothy. *Carbon Democracy: Political Power in the Age of Oil*. Verso, 2011.

Mitman, Gregg. "Hubris or Humility: Genealogies of the Anthropocene." In *Future Remains: A Cabinet of Curiosities for the Anthropocene*, ed. Gregg Mitman, Marco Armiero, and Robert S. Emmett. University of Chicago Press, 2017.

Mitman, Gregg, Marco Armiero, and Robert Emmett. Future Remains: A Cabinet of Curiosities for the Anthropocene. University of Chicago Press, 2017.

Mitman, Gregg, Michelle Murphy, and Christopher Sellers. "Introduction: A Cloud over History." *Osiris* 19 (2004): 1–17.

Möllers, N., C. Schwägerl, and H. Trischler, eds. *Welcome to the Anthropocene: The Earth in Our Hands*. Deutsches Museum / Rachel Carson Center, 2014.

Moore, Jason W. *Capitalism in the Web of Life*. Verso, 2015.

Moore, Kathleen Dean, and Michael Nelson. *More Ground: Ethical Action for a Planet in Peril*. Trinity University Press, 2011.

Moore, Stephen. *Alternative Routes to the Sustainable City*. Lexington Books, 2007.

Morton, Timothy. *The Ecological Thought*. Harvard University Press, 2010.

Morton, Timothy. "Guest Column: Queer Ecology." *PMLA* 273–282.

Mumford, Lewis. *The Myth of the Machine*. Harcourt Brace Jovanovich, 1979.

Myers, Natasha. *Rendering Life Molecular*. Duke University Press, 2015.

Neihardt, John. *Black Elk Speaks: Being the Life Story of a Holy Man of the Oglala Sioux*. University of Nebraska Press, 1979.

Nestle, Marion. *Food Politics: How the Food Industry Influences Nutrition and Health*, revised edition. University of California Press, 2007.

Nixon, Rob. "The Anthropocene: Promises and Pitfalls of an Epochal Idea." *Edge Effects*, posted November 6, 2014. http://edgeeffects.net/anthropocene-promise -and-pitfalls/

Nixon, Rob. *Slow Violence and the Environmentalism of the Poor*. Harvard University Press, 2011.

Nye, David E. *America as Second Creation: Technology and Narratives of New Beginnings*. MIT Press, 2003.

Nye, David E. *America's Assembly Line*. MIT Press, 2013.

Nye, David E. *Consuming Power*. MIT Press, 1998.

Nye, David E. *Technology Matters: Questions to Live With*. MIT Press, 2006.

Nye, David E., ed. *Technologies of Landscape: Reaping to Recycling*. University of Massachusetts Press, 1999.

Nye, David E. *When the Lights Went Out: A History of Blackouts in America*. MIT Press, 2010.

Nye, David E., Linda Rugg, James Fleming, and Robert Emmett. "The Emergence of the Environmental Humanities." Stockholm: Mistra, May 2013. http://www.mistra .org/download/18.7331038f13e40191ba5a23/Mistra_Environmental_Humanities_ May2013.pdf

Ohl, Michael. *Die Kunst der Benennung*. Matthes & Seitz, 2015.

Oreskes, Naomi, and Erik M. Conway. *The Collapse of Western Civilization: A View from the Future*. Columbia University Press, 2014.

Oreskes, Naomi, and Erik M. Conway. *Merchants of Doubt*. Bloomsbury, 2010.

Orr, David W. *Down to the Wire: Confronting Climate Collapse*. Oxford University Press, 2009.

Owen, David. *Green Metropolis*. New York: Riverhead Books, 2009.

Parr, Adrian. *Hijacking Sustainability*. MIT Press, 2009.

Pellow, David Naguib. *Resisting Global Toxics: Transnational Movements for Environmental Justice*. MIT Press, 2007.

Phillips, Dana. *The Truth of Ecology*. Oxford University Press, 2003.

Piketty, Thomas. *Capital in the Twenty-first Century*, trans. A. Goldhammer. Harvard University Press, 2014.

Plumwood, Val. "Surviving a Crocodile Attack." *Utne Reader*, July-August 2000. http://www.utne.com/arts/being-prey

Plumwood, Val. *The Eye of the Crocodile*, ed. L. Shannon. Australian National University Press, 2012.

Pratt, Joseph, and Martin Melosi. "The Energy Capital of the World? Oil-Led Development in Twentieth Century Houston." In *Energy Capitals: Local Impact, Global Influence*, ed. Pratt and Melosi. University of Pittsburgh Press, 2014.

Princen, Tom, and Michael Maniates. *Confronting Consumption*. MIT Press, 2002.

Purdy, Jedediah. *After Nature: A Politics for the Anthropocene*. Harvard University Press, 2015.

Purdy, Jedediah. "Anthropocene Fever." *Aeon*, March 31, 2015.

Radermacher, Walter. "Indicators, Green Accounting and Environment Statistics: Information Requirements for Sustainable Development." *International Statistical Review* 67, no. 3 (1999): 339–354.

Radkau, Joachim. *Nature and Power: A Global History of the Environment*. Cambridge University Press, 2008.

Reece, Erik. *Lost Mountain: A Year in a Vanishing Wilderness*. Riverhead Books, 2006.

Regalado, Antonio. "Ethical Questions Loom Over Efforts to Make a Human Genome from Scratch." *MIT Technology Review*, May 25, 2016.

Regalado, Antonio. "Paint-On GMOs Could Create Cattle, Dogs, with Custom Fur." *MIT Technology Review*, May 2016, 17.

Regan, Tom. *The Case for Animal Rights*. Updated edition. University of California Press, 2004.

Rigby, Kate. "Spirits that Matter: Pathways towards a Rematerialization of Religion and Spirituality." In *Material Ecocriticism*, ed. Serenella Iovino and Serpil Oppermann. 283–290. Indiana University Press, 2014.

Roberts, Jody A., and Nancy Langston. "Toxic Bodies/Toxic Environments: An Interdisciplinary Forum [Introduction]." *Environmental History* 13, no. 4 (2008): 629–635.

Robin, Libby. "New Science for Sustainability in an Ancient Land." In *Nature's End: History and the Environment*, ed. Sverker Sörlin and Paul Warde. Palgrave MacMillan, 2009.

Robin, L., S. Sörlin, and P. Warde, eds. *The Future of Nature*. Yale University Press, 2013.

Robinson, John. "Being Undisciplined: Some Transgressions and Intersection in Academia and Beyond." *Futures* 40 (2008): 70–86.

Rockström, Johan, et al. "A Safe Operating Space for Humanity." *Nature* 472–475.

Rodney, Walter. *How Europe Underdeveloped Africa*. London: Bogle-L'Ouverture, 1972.

Rose, Deborah Bird. "Val Plumwood's Philosophical Animism: Attentive Interactions in the Sentient World." *Environmental Humanities* 3, 2013: 93–109.

Sale, Peter F. *Our Dying Planet: An Ecologist's View of the Crisis We Face*. University of California Press, 2011.

Salgado, Sebastião. *Genesis*. Taschen Books, 2013.

Sandilands, Catriona. "Sexual Politics and Environmental Justice." In *New Perspectives on Environmental Justice: Gender, Sexuality and Activism*, ed. Rachel Stein, 109–126. Rutgers University Press, 2004.

Scarry, C. Margaret. "Patterns of Wild Plant Utilization." In *People and Plants in Ancient Eastern North America*, ed. Paul E. Minnis, 50–104. Smithsonian Books, 2003.

Schudsen, Michael. *Advertising: The Uneasy Persuasion*. Basic Books, 1984.

Scranton, Roy. *Learning to Die in the Anthropocene: Reflections on the End of Civilization*. City Lights, 2015.

Segal, Howard P. *Utopias: A Brief History from Ancient Writings to Virtual Communities*. Wiley, 2012.

Sekula, Alan. *Fish Story*. Witte de With and Richter, 1995.

Self, Robin M., Donald Self, and Janel Bell-Haynes. "Marketing Tourism in the Galapagos Islands: Ecotourism or Greenwashing?" *International Business and Economics Research Journal* 9, no. 6 (2010): 111–125.

Serpas, Martha. "Corollary." In *The Dirty Side of the Storm*. Norton, 2007.

Seymour, Nicole. *Strange Natures: Futurity, Empathy, and the Queer Ecological Imagination*. Illinois University Press, 2013.

Sideris, Lisa. "Anthropocene Convergences: A Report from the Field." *RCC Perspectives*, no. 2, 2016: 89–96.

Singer, Peter. *Animal Liberation*. Harper Perennial, 2009.

Singer, Peter, ed. *In Defense of Animals*. Blackwell, 1985.

Sloterdijk, Peter. The Anthropocene: A Process-State on the Edge of Geohistory? In *Textures of the Anthropocene: Grain, Vapor, Ray*, volume III, ed. Katrin Klingan, Ashkan Sepahvand, Christoph Rosol, and Bernd M. Scherer, 257–271. MIT Press, 2015.

Snyder, Gary. *The Practice of the Wild*. Milkweed, 1990.

Sörlin, Sverker. "Environmental Humanities: Why Should Biologists Interested in the Environment Take the Humanities Seriously?" *Bioscience* 62, no. 9 (2012): 788.

Sovacool, Benjamin. *The Dirty Energy Dilemma: What's Blocking Clean Power in the United States*? Praeger, 2008.

Stasch, Rupert. "Toward Symmetric Treatment of Imaginaries: Nudity and Payment in Tourism to Papua's 'Treehouse People.'" In *Tourism Imaginaries: Anthropological Approaches*, ed. Noel B. Salazar and Nelson H. H. Graban. Berghahn, 2014.

Stein, R., ed. *New Perspectives on Environmental Justice: Gender, Sexuality, and Activism*. Rutgers University Press, 2004.

Steinberg, Ted. *Down to Earth: Nature's Role in American History*. Oxford University Press, 2002.

Stengers, Isabelle. *Au Temps des Catastrophes: Résister à la Barbarie Qui Vient*. Editions la Découverte, 2013.

Stiglitz, Joseph. *The Roaring '90s*. Norton, 2003.

Stoner, Alexander M., and Andony Melathopoulos. *Freedom in the Anthropocene: Twentieth-Century Helplessness in the Face of Climate Change*. Palgrave-Macmillan, 2015.

Storm, Anna. *Post-Industrial Landscape Scars*. Palgrave-Macmillan, 2014.

Strasser, Susan. *Satisfaction Guaranteed: The Making of the American Mass Market*. Pantheon, 1989.

Suess, Eduard. *The Face of the Earth*, volume 2, trans. Hertha B. C. Sollas. Clarendon, 1906.

Sullivan, Heather. "Dirt Theory and Material Ecocriticism." *Interdisciplinary Studies of Literature and Environment* 19, no. 3 (2012): 515–531.

Sullivan, Jonathan, and Lei Xie. "Environmental Activism, Social Networks and the Internet." *China Quarterly* 198 (2009): 422–432.

Tall, Deborah. *From Where We Stand: Recovering a Sense of Place*. Johns Hopkins University Press, 1993.

TallBear, Kimberly. "Shepard Krech's *The Ecological Indian*, One Indian's Perspective." *International Institute for Indigenous Resource Management* (September 2000): 1–6. http://www.iiirm.org/publications/Book%20Reviews/Reviews/Krech001.pdf

Tester, Jefferson W., Elisabeth M. Drake, Michael J. Driscoll, Michael W. Golay, and William A. Peters. *Sustainable Energy: Choosing Among Options.* MIT Press, 2005.

Thomashow, Mitchell. *Ecological Identity.* MIT Press, 1995.

Thoreau, Henry David. *Walden* (1854). Ed. J. Lyndon Shanley. Princeton University Press, 1971.

Thoreau, Henry David. *A Week on the Concord and Merrimack Rivers* (1849). Penguin, 1998.

Traer, Robert. *Doing Environmental Ethics.* Westview, 2009.

Trentmann, Frank. *Empire of Things: How We Became a World of Consumers, from the Fifteenth Century to the Twenty-first.* Penguin Books, 2016.

Tuan, Yi-Fu. *Space and Place: The Perspective of Experience.* Minnesota University Press, 1977.

Tufnell, Ben. *Land Art.* Tate, 2006.

Urry, John. *The Tourist Gaze.* SAGE, 1990.

van Dieren, Wouter, ed. *Taking Nature into Account: A Report to the Club of Rome-Toward a Sustainable National Income.* Springer, 1995.

van Dooren, Thom. *Flight Ways: Life and Loss at the Edge of Extinction.* Columbia University Press, 2014.

Walker, Brett L. *The Lost Wolves of Japan.* University of Washington Press, 2005.

Wall, Derek. *The Commons in History.* MIT Press, 2014.

Walpole, M. J., G. G. Karanja, N. W. Sitati, and N. Leader-Williams. *Wildlife and People: Conflict and Conservation in Masai Mara.* Kenya: International Institute for Environment and Development, 2003.

Wane, Njoki Nathani. *Indigenous African Knowledge Production: Food-processing Practices among Kenyan Rural Women.* University of Toronto Press, 2014.

Watson, Janell. "Eco-Sensibilities: An Interview with Jane Bennett." *Minnesota Review* 81 (2013): 147–148.

Watts, Alan. *"The World Is Your Body."* In *The Ecological Conscience: Values for Survival*, ed. Robert Disch. Prentice-Hall, 1970.

Watts, M., and R. Peet, eds. *Liberation Ecologies: Environment, Development, Social Movements.* Routledge, 2004.

Weik von Mossner, Alexa. "Science Fiction and the Risks of the Anthropocene: Anticipated Transformations in Dale Pendell's *The Great Bay.*" *Environmental Humanities* 5 (2014): 203–216.

Weiss, Monica. *The Environmental Vision of Thomas Merton.* University Press of Kentucky, 2011.

West, Rinda. *Out of the Shadow: Ecopsychology, Story, and Encounters with the Land.* University of Virginia Press, 2007.

Weyler, Rex. "Nature's Apprentice: A Meta-Narrative for Aging Empires." *Manoa* 25, no. 1 (2013): 187–196.

White, Leslie. *Science of Culture.* Grove, 1949.

Whitman, Walt. *Leaves of Grass,* ed. Sculley Bradley and Harold W. Blodgett. Norton, 1973.

Williams, Raymond. "Ideas of Nature." In *Problems in Materialism and Culture.* Verso, 1980. 67–85.

Wilson, E. O., and M. Frances Peter. *Biodiversity.* National Academy Press, 1988.

Winkler, Allan M. *Life Under a Cloud: American Anxiety about the Atom.* Oxford University Press, 1993.

Wooley, Agnes. "'There's a Storm Coming:' Reading the Threat of Climate Change in Jeff Nichols's *Take Shelter.*" *Interdisciplinary Studies in Literature and Environment* 21, no. 1 (2014): 174–191.

Zalasiewicz, Jan, et al. "The Technofossil Record of Humans." *Anthropocene Review* 1, no. 1 (2014): 34.

Zalasiewicz, Jan. "Is Earth in a New Geological Phase Thanks to Us?" *New Scientist,* November 2014: 5.

Ziolkowsi, Lori A. "The Geologic Challenge of the Anthropocene." *RCC Perspectives,* no. 2, 2016: 35–39.

Žižek, Slavoj. *Living in the End Times.* Verso, 2010.

Zube, Ervin H. *Landscapes: Selected Writings of J. B. Jackson.* University of Massachusetts Press, 1970.

Zylinska, Joanna. *Minimal Ethics of the Anthropocene.* Open Humanities Press, 2014.

Index

Abbey, Edward, 3, 28
Abram, David, 160
Abu Dhabi, 66
Acid rain, 8, 48, 84, 97, 102
Actor-network theory, 143
Adams, Ansel, 22f, 25, 43
Adamson, Joni, 17, 149
Addams, Jane, 3
Adjaye, David, 92, 107
Advertising, 56–59, 124, 185n15. *See also* Consumption.
Africa, 13, 15–16, 18–19, 25, 45, 50, 62, 85, 100, 106, 147–149. *See also* specific countries.
  degrowth and, 124
  drought and, 171
  localization and, 119
  mining and, 96, 112
  national parks and, 13, 34
  nature reserves and, 40–41
  wilderness and, 40, 43
Agnoletti, Mauro, 83–84
Agriculture, 11–12, 15–16, 26, 83–85, 96–98, 131, 172. *See also* Gardens.
  Africa and, 106
  Australia and, 119–120
  biodiversity and, 83–84
  extinction and, 96–97, 168
  fertilizers and, 84–85, 89, 155
  First Peoples and, 149
  industrialization of, 39, 59, 99

  localization and, 119–122
  new materialism and, 142, 145
  plantations and, 15–16, 98, 174
  soil depletion and, 26, 112
  Thoreau and, 27–28
  urban, 65, 124, 172, 186n36
Agyeman, Julian, 66
Air pollution, 1, 35, 51, 55, 57, 86, 105, 128, 173
Alaimo, Stacy, 129, 156
Albedo modification, 87
Albrecht, Glenn, 112
Alexievich, Svetlana, 109
Allen, Barbara, 172
Allen, Durward, 37
Alvear, Cecilia, 36
Amazonia, 16, 39, 43, 98, 150, 166, 171
American Society for Environmental History, 17
American West, 15, 19, 39
Animals. 7, 13–15, 35, 38, 41–42, 78–84, 152–156, 176. *See also* Endangered species, Extinction, Hunting.
  cats, 14, 155
  cattle, 15, 19, 155, 183n45
  co-becoming and, 14, 102
  companion species and, 14, 21, 175
  consciousness and, 153, 176
  dogs, 14, 19, 77, 155
  domestic, 6, 12, 14–15, 26, 38–39, 47, 153, 155

Animals (cont.)
  First Peoples and, 153–154
  horses, 15, 39, 62,
  nature reserves and, 13, 35, 38, 40
  preservation and, 13–19, 22, 32,
    35–36, 39, 42, 148, 175
  queer ecology and, 152, 160
  World Wildlife Fund (WWF) and, 121,
    166–167
Antarctic Sea, 85
Anthropocene, 16–17, 93–114. *See also*
    Extinction, Global warming
  actual disasters and, 109–114
  apocalyptic narratives and, 104–109
  Crutzen and, 95, 98, 102, 166
  Dark Mountain Project and, 106–107
  disasters and, 93, 95, 100, 108–111
  environmental education and, 95,
    104–109
  ethics and, 93–95
  geoengineering and, 95–96, 175
  localization and, 118–119
  Mass Extinction Memorial Observa-
    tory and, 92f, 107–108
Anthropology, 3, 5–6, 8, 12, 14, 21,
    36, 170
  animals and, 152, 154
  biodiversity and, 80, 83
  biotechnology and, 74, 76, 78
  consumption and, 47, 54–55, 62, 68
  ecological economics and, 139–140
  ecotourism and, 36
  First Peoples and, 145–146, 151. *See
    also* First Peoples
  geoengineering and, 88
  new materialism and, 140
  queer ecology and, 157
  restructuring knowledge and, 164
  wilderness and, 38–39, 43
Appliances, 20, 49, 51–55, 58, 64,
    166–167
Aral Sea, 84
Architecture for Humanity, 134

Arctic Ocean, 84
Arroyo Seco Confluence, 116f
Asia. *See* China, India, Indonesia, Japan,
    Sri Lanka.
Astrobiology, 75, 187n10
Atomic power, 12, 48, 96, 104–105,
    109, 149
Atwood, Margaret, 77, 138f, 156
Audi, 57
Austin, Texas, 67
Australia, 5, 112, 119–121
  animals and, 154
  coal and, 52
  consumption levels in, 50
  ecotourism and, 36
  environmental humanities and, 3–6,
    8, 15, 17–18
Australian Western Mining Corpora-
    tion, 117
Automobiles, 49–52, 57, 61, 68, 122,
    126–127
  degrowth and, 124–126
  ecological economics and, 131, 133
  electric, 49, 131, 139, 168–169
  fuel efficiency (mpg) of, 49, 52
  gasoline and, 48–49, 52, 57, 65, 131,
    167–168recycling and, 127, 129
  sustainable cities and, 65–66
Avatar (film), 90
Aztecs, 12, 26
Azzarello, Robert, 158

Bacteria, 75–76
Baichwal, Jennifer, 111–112
Barad, Karen, 140, 143
Barrow, Mark, 79
Bartram, William, 150
Bateson, Gregory, 83, 152
Beaver Lake Cree nation, 134, 147
Beck, Ulrich, 105
Belize, 35
Bengal, 19
Benjamin, Walter, 112

Bennett, Jane, 140–144
Bentham, Jeremy, 155
Benzene, 172–173
Berry, Wendell, 62, 118
Bertoni, Filippo, 187n10
Bhopal, 105
Bhutan, 36
Bicycles, 30, 36, 61, 65, 121–122, 165,
  169
Bill and Melinda Gates Foundation, 88
Biodiversity, 42, 78–84
  Abney Park and, 159
  degrowth and, 124
  ecotourism and, 31, 35, 81
  endangered species and, 7, 15, 31,
    35–36, 79, 83, 99, 139, 154, 173, 175
  forests and, 29, 80–83
  place and, 25, 29, 31, 35, 38, 45
  preservation and, 25, 29, 80, 83
Bioeconomy, 73
Biomimetics, 126–127, 129, 139, 176
Biopolitics, 75–76
Biotechnology, 72–78
  animals and, 155–156
  DNA and, 5, 72–74, 77, 89–91, 153,
    155–156, 175
  environmental art and, 74–75
  genetics and, 8, 11, 72–74, 77–78, 89,
    104, 117, 154, 156
  place and, 71–78
  transgenic species and, 77, 156
Birds, 2, 14, 15, 26, 41, 64, 74, 78–79,
    81, 143, 154, 155
  cranes, 83
  Dodo, 92f
  falcons, 14,
  Great Auk, 1, 15
  passenger pigeons, 74
Birth rates, 100, 123
Black Elk, 11, 154
Blake, William, 105
Bolivia, 17, 89, 173–174
Bolster, Jeffrey, 135

Bonpland, Aimé, 4
Bookchin, Murray, 99
Botswana, 32
BP, 105
Braidotti, Rosi, 140, 143
BRAIN (Brain Research through Advanc-
  ing Innovative Neurotechnologies),
  72–73, 186n4
Brand, Stewart, 103
Braungart, Michael, 126–127
Brazil, 17, 41–44, 53, 84, 96, 120
Britain, 6, 13, 16, 29, 52, 54, 120, 125,
  129
Brokeback Mountain (film), 159
Brueghel, Pieter the Elder, 108
Buddhism, 11, 35, 45, 62–63
Buell, Frederick, 111
Buell, Lawrence, 109, 118
Building codes, 65, 68
Burtynsky, Edward, 111–112
Bush, George W., 131
Buzard, James, 32

Callicott, J. Baird, 150–151
Calvin, William H., 100
Canada, 78, 169
  consumption and, 50, 59
  First Peoples and, 134, 147, 149
  hydroelectric resources of, 52
  salmon and, 165–166
  wolves and, 37
Canals, 63, 81, 84, 112
Cancer
  biodiversity and, 82
  biotechnology and, 75, 77
  chemicals and, 17, 167, 172
  detection of by animals, 155
  ecocide and, 109–110
  ecological economics and, 134
  industry and, 167, 172, 189n3
  pollution and, 17, 109–111, 167, 172,
    189n3
  smoking and, 167

Can Decreix, 123
Cape Cod, 30
Capitalism, 24, 56–59, 75, 94, 105–106, 111, 124, 131–134. *See* Fossil Fuels, Industry.
    First Peoples and, 145
    global warming and, 103
    green imagery and, 139
    new materialism and, 142
    queer ecology and, 158
Carbon, 8, 20, 28, 31–32, 49, 50–54, 59–61, 64–66, 94, 100, 121, 123, 130, 132–134, 136, 139, 166
    cattle methane and, 155
    deforestation and, 155
    geoengineering and, 85, 87–88
    industrial revolution and, 97
    oceans and, 85, 95–97
    pre-Columbian America and, 98
    rural vs. city dwellers and, 28
    taxes and, 168
Carpenter, Novella, 29
Carrigan, Anthony, 34
Carson, Rachel, 3, 26–27, 63, 99, 105, 109, 162f, 167
Cartesian dualism, 142, 154–155, 164
Catholicism, 10, 17
Cayuga, 25
Cell phones, 11, 62
Center for Sustainable Landscapes (CSL), 128
CFCs, 99, 166
Chakrabarty, Dipesh, 97, 102, 104
Cheap nature, 142, 169
Chemicals *See also* Pesticides, Petro-chemicals, Toxins
    animals and, 155
    bacteria and, 75
    banned, 18
    Bhopal and, 105
    cancer and, 17, 167, 172
    ecological economics and, 133
    fertilizers and, 84–85, 89, 155

First Peoples and, 146, 149
    geoengineering and, 88
    leaching agents and, 75–76
    new materialism and, 144
    pollution and, 99, 105, 109, 172–173
    regulation and, 18
Cheney, Jim, 146
Chernobyl, 34, 48, 105, 109–111
Chicago School, 65
China, 19, 53, 124, 167
    aristocratic hunting parks and, 44
    carbon emissions of, 53, 54
    consumption level of, 59
    exporting toxins to, 18, 62, 112
    green cities and, 2, 127
    mining and, 96
    one-child-per-couple policy of, 51, 53
Christians, 10–11, 40, 43, 62–63, 184n55
Church, George, 73
Cities, 26, 28–29, 61–69, 105, 121–122
    cosmopolitanism and, 29, 108, 136
    cycling, 68, 133, 186n43
    ecological economics and, 2, 130–131, 133, 136
    evacuation of, 105
    future, 47, 170
    green, 2, 28, 47, 68, 121–122, 126–128, 133, 139, 186n43
    redeveloping, 175
    restructuring knowledge and, 164
    Roman Empire and, 26
    street life and, 169
    water supply and, 171
Citizen's Clearing House for Hazardous Wastes, 17
Civil rights, 17
Climate change, 3–8, 37–38, 49–50, 54–56, 142, 160, 166, 168–169. *See also* Consumption, Energy
    Anthropocene and, 93–97, 99, 101–104, 106, 108
    biodiversity and, 82

capitalism and, 103
degrowth and, 123, 125
denial of, 131, 176
ecological economics and, 131–134, 136
environmental humanities and, 1,3–4, 7–8, 16
excessive academic specialization and, 164
geoengineering and, 84–87, 89
greenhouse gases and, 49, 55, 87, 99, 136, 175
Intergovernmental Panel on Climate Change and, 93
localization and, 122
People's World Conference on Climate Change and the Rights of Mother Earth and, 89, 174
sacrifice zones and, 18, 172–174
Climate justice, 10, 16–17
Club of Rome, 105, 123
Coal
Anthropocene and, 96, 104, 112
consumption of, 4, 48–49, 52, 55, 57, 96, 104, 112, 122, 125, 156, 167
corporate lobbies for, 167
degrowth and, 125
First Peoples and, 39
Jevon on, 4
localization and, 122
Co-becoming, 14, 102
Cocker, Mark, 29
Cohen, Lizabeth, 56–57
Cold War, 86, 109
Commodification, 94, 129–130, 199
Commodity regionalism, 117, 128–131, 176
Companion species, 14, 21, 175
Computers, 18, 52, 71, 90, 140, 166
Conservation
Anthropocene and, 106
biodiversity and, 78–82
capitalism and, 106

crisis of the commons and, 12–14
ecotourism and, 33–34
environmental justice and, 17
First Peoples and, 18–19, 148–151
focus on energy, 175
forests and, 30
wildlands and, 38, 40–41
Consumption, 20, 56–64, 74. *See also* Degrowth, Ecological footprint, Energy, Food
advertising and, 56–59, 124, 185n15
Anthropocene and, 111
anthropology and, 47, 54–55, 59, 62, 68
China and, 59
climate change and, 60
coal and, 4, 48–49, 52, 55, 57, 96, 104, 112, 122, 125, 156, 167
electricity and, 52–57, 60, 64–68
energy and, 48–57
ethics and, 50, 111
First Peoples and, 147–148
grassroots organizations and, 59, 167
greenwashing and, 60, 68, 164, 167
localization and, 120, 123
sustainability and, 19–20, 50–56, 59–60, 66–69, 127
waste and, 62, 64–68, 127 (*see also* Waste)
Contact zones, 25, 62
Conway, Erik, 103
Coral reefs, 15, 45, 93, 114, 150
Cosmopolitanism, 29, 108, 136
Costa Rica, 32, 36
Coté, Charlotte, 150
Courtney, Jason, 138f
Cradle-to-cradle design, 126–129. 133–135
Crisis of the commons, 12–14, 21, 175
Crist, Eileen, 103–104
Cronon, William, 40, 129
Crosby, Alfred W., 15
Crutzen, Paul, 85–86, 95, 98, 102, 166

Cuba, 122
Cushman, Greg, 98
Cyborgs, 11, 90, 102, 140, 175
Cycling cities, 65, 68, 133, 169, 186n43

Dams, 26, 78, 81, 84, 99, 119
Daniels, Gene, 22f, 70f
Dark Mountain Project, 106–107
Darwin, Charles, 80, 153
Dauvergne, Peter, 124
DDT, 99
Deakin, Roger, 29
De Chardin, Pierre Teilhard, 99
Deepwater Horizon, 109
Deforestation, 19, 24, 26, 96–100, 124, 168
Degrowth, 20–21, 123–127, 139. *See also* Sustainability
  ecological economics and, 131, 133
  internal challenges and, 171, 176
  water and, 126, 169
Deleuze, Gilles, 140
DeLillo, Don, 31
Denes, Agnes, 130
Denmark, 55, 64–65, 121–122
Deserts, 44–45, 66, 84–85, 99, 168, 175
Deutch, John, 54
De Young, Raymond, 118–119
Disease, 20, 72, 76, 105, 117, 119
DNA, 5, 72–74, 77, 89–91, 153, 155–156, 175
Dooling, Sarah, 66
Down-cycling, 126
Drought, 20, 72, 93, 100, 112, 171
Drury, Chris, 129–130

Earth First!, 165
Ecocide, 109–111
Ecocriticism
  alienation and, 146–147
  apocalyptic narratives, 104–107
  biodiversity and, 82
  climate fictions and, 108, 110

cosmopolitanism and, 29, 108, 136
dystopias and, 77, 93, 156
ecotourism and, 34
environmental humanities and, 6, 8, 14, 19, 170
First Peoples and, 145–146
interdisciplinary research and, 170–171
material, 144–145
queer ecology and, 156–159
Ecological citizenship, 6, 136, 167, 199n67
Ecological education, 93–95, 114. *See also* Environmental education
Ecological economics, 8, 130–136, 171, 176
biotechnology and, 73
cities and, 2, 64, 130–131, 133, 136
climate change and, 131, 136
dams and, 119
ecocide and, 109–111
fossil fuels and, 120, 126, 133, 135
India and, 17
patents and, 71, 75, 77, 88
renewable energy and, 50, 132, 136
waste and, 133, 139
Ecological footprint, 133
Ecological gentrification, 66
Ecological imperialism, 6, 15–16, 175
Ecological rift, 105, 142
Ecopsychology, 145–146
Ecoracism, 4, 18–19, 21, 175. *See also* Environmental justice
Ecosystem services, 80, 82
Ecotourism, 30–37, 44, 167
biodiversity and, 31, 35, 81
fossil fuels and, 167
industry and, 31–32, 34, 47, 167
internal challenges and, 175
International Ecotourism Society and, 33
wilderness and, 175
Ecovillages, 2, 119–121, 131

Ecuador, 36, 39, 147, 173
Edelman, Lee, 158
Eiseley, Loren, 145
Electricity. *See also* Gas, Solar energy,
    Wind power
  appliances and, 20, 49, 51–55, 58, 64,
    166–167
  biotechnology and, 77
  blackouts and, 143
  buses and, 64
  cars and, 49, 131, 139, 168–169
  consumption and, 52–57, 60, 64–68
  degrowth and, 125
  ecological economics and, 131
  geoengineering and, 84
  localization and, 119–123
  new materialism and, 143
  pollution and, 64
  recycling and, 128
Elephants, 21, 154
Emerson, Ralph Waldo, 106–107
Endangered species, 7, 15, 31, 35–36,
    79, 99, 139, 150, 154, 173, 175
Ende Gelände, 125
"End of Life Vehicle Directive" (Euro-
    pean Union), 127
Energy, 47–56, 167, 175. *See also* Coal,
    Electricity, Gas, Solar energy, Wind
    power
  atomic, 12, 48, 96, 104–105, 109, 149
  automobiles and, 49–52
  building codes and, 65, 68
  carbon and, 20, 50–54, 139
  consumers and, 1, 50, 55–56, 66
  consumption and, 6, 49–52, 55–56 (*see
    also* Consumption)
  efficiency and, 1, 20, 33, 49–51, 69,
    122, 124, 126
  fossil fuels and, 4, 48–50, 52, 55,
    57–58, 65, 66, 69, 81, 99, 104, 112,
    120–122, 125, 131 133, 135, 155,
    167–168, 175
  geothermal, 46f, 121, 128

  industry and, 48–49, 52, 55, 104
  Kaya Identity and, 51–54, 63, 69, 123
  nuclear, 8, 34, 48–49, 52, 84, 95, 100,
    105, 109–111, 149, 173
  petrochemicals and, 17, 172–173
  renewable, 55, 122, 133, 166
  steam power and, 47, 64
Enlightenment, 10, 90, 102
Environmental art, 74–75, 106–107,
    128–131
Environmental despair, 106
Environmental education, 60, 93–95,
    104, 114, 120–121, 149
Environmental history, 8, 14. *See also*
    Anthropocene.
  American Society for Environmental
    History and, 17
  Anthropocene and, 16, 98
  ecoracism and, 18
  ecotourism and, 30–31, 35
  energy and, 47–48
  India and, 29
  Marsh and, 3–4
  oceans and, 25, 95, 97, 135
  place and, 25–27, 29–30, 35
  universities and, 6, 70–171
Environmental humanities. *See also* An-
    thropology, Ecocriticism, Environ-
    mental History, Ethics
  Anthropocene and, 93–114
  biotechnology and, 72–78
  central concepts of, 8–21
  consumer choice and, 59
  ecotourism and, 31–37, 44, 81, 175
  ethics and, 4–6, 10, 44, 93–94
  internal challenges to, 170–176
  opportunities in, 165–170
  place and, 23–30
  postcolonial studies and, 4, 6, 8,
    18–19, 34, 140, 145–147
  restructuring knowledge and, 163–165
Environmental justice, 4, 17–18,
    148–149. *See also* Ecoracism.

Environmental justice (cont.)
  biotechnology and, 77
  crisis of the commons and, 12–14, 21,
    175
  First Peoples and, 148–149
  internal challenges and, 172
  New Orleans and, 107
  queer ecology and, 157–159
  regulation and, 168
  sacrifice zones and, 18, 172–174
Epistemology, 8–9, 76, 102, 109, 140,
  146
Erosion, 100, 113
Ethics
  animals and, 38, 152–155
  Anthropocene and, 93–95
  biodiversity and, 82–83
  biotechnology and, 72–74, 77–78,
    156
  communitarian, 168–169
  consumption and, 50
  ecotourism and, 34
  environmental humanities and, 44,
    93–94, 171, 173
  First Peoples and, 145–149, 151
  geoengineering and, 85–88
  inventions and, 73
  land use and, 93
  new materialism and, 142, 144
  queer ecology and, 158–159
  restructuring knowledge and, 163
  science and, 101–102
Ethnography. See also Anthropology.
  biodiversity and, 80, 83
  biotechnology and, 76
  consumption and, 59, 68
  contact zones and, 25, 62
  First Peoples and, 145, 151–152
  multispecies, 39, 83, 145
Europe, 18–19, 39, 47–50, 54–57, 125,
  129. See also specific countries.
  Anthropocene and, 98
  consumption and, 49–50, 59

  ecotourism and, 35
  imperialism and, 15–16, 30, 47, 104,
    147
  emergence of environmental humani-
    ties and, 3–4, 132
  lower carbon development and, 54–55
  wilds and, 34, 37–40, 89
European Union, 18, 35, 55, 127, 129,
  169
Evolution. See also DNA, Extinction
  Africa and, 25
  animals and, 153
  Anthropocene and, 95, 99
  biodiversity and, 80, 84
  cyborgs and, 10, 175
  ecocritical reading and, 145
  environmental humanities and, 10,
    12, 14
  First Peoples and, 150
  queer ecology and, 160
  social, 48
E-waste, 18, 62, 112, 166
Extinction, 1, 3, 11–12, 19–20, 78–84,
    169. See also Animals, Birds, Endan-
    gered species, Hunting
  agriculture and, 96–97, 168
  animal rights and, 11, 19–20, 24, 36,
    38
  Anthropocene and, 91, 93–99,
    107–108
  biotechnology and, 72–74
  degrowth and, 123
  ecological economics and, 133, 136
  ecotourism and, 36
  excessive specialization and, 164
  First Peoples and, 150, 166
  human, 175–176
  Mass Extinction Memorial Observa-
    tory and, 92f, 107–108
  new materialism and, 145
  place and, 24, 36, 38, 41
  sixth, 14–15, 96, 175–176
Exxon, 55, 109, 131

Farley, Paul, 29
Fear the Walking Dead (series), 130
Feinberg, Leslie, 159
Feminism, 4, 8, 139–140, 146–147, 149, 156–157, 173, 178
Fertilizers, 84–85, 89, 155
Fiennes, William, 29
First Peoples. 13–14, 18–19, 23, 39, 48, 145–154, 166. *See also* specific peoples
  biodiversity and, 82
  conservation and, 18–19, 148–151
  ecological economics and, 134
  forests and, 16–17, 39, 148–150
  hunting and, 11, 23–24, 35, 148–151
  InterTribal Sinkyone Wilderness Council and, 149
  mining and, 147, 149, 197n34
  oil and, 19, 147, 149–150
  People's World Conference on Climate Change and the Rights of Mother Earth and, 89
  sacred things and, 147, 149–150
  toxins and, 134, 149–150
  UN Indigenous People's Global Summit on Climate Change and, 89
  water and, 24, 44, 148–149
  wilderness and, 24, 98, 149
  wind energy and, 62
Fishing, 13, 62, 82, 96, 98, 122, 130–131, 135, 150, 165, f179
Fleming, James Rodger, 86–88
Flying squirrels, 113–114
Food, 7, 9, 58–61, 64, 66, 81, 100, 106, 117, 120–124, 155, 167, 175. *See also* Agriculture, Animals, Fishing, Gardens, Hunting, Oceans.
  Anthropocene and, 94, 100, 109–110, 112
  degrowth and, 123–124
  ecological economics and, 130
  ecological imperialism and, 6, 15–16, 175

ecotourism and, 31, 35
First Peoples and, 145, 150
greenhouses and, 122, 124
hunting and, 11–12, 15, 19, 23–24, 35, 39–41, 44, 47, 74, 82, 134, 148–151, 169, 183nn45,51
new materialism and, 143
pollution and, 26, 109
radioactive, 110
refrigerators and, 10, 51, 54–55, 64, 166
Slow Food movement and, 167
toxins and, 109, 134
Ford Motor Company, 126
Forests, 14, 16–17, 19, 24, 26, 36, 29–30, 80–83, 96–100, 112, 114, 124, 168, 173
  animals and, 41–42, 155
  Anthropocene and, 93, 96–100, 106, 112–113
  biotechnology and, 74
  birdsong and, 41
  ecological economics and, 132, 135
  ecotourism and, 36
  First Peoples and, 16–17, 39, 40, 148–150
  food and, 120
  localization and, 119–120
  rain, 24, 80–82, 93, 106, 135
  reforestation and, 36, 39, 42, 83, 98
  religion and, 41, 44
Fossil fuels. *See also* Coal, Gas, Oil
  animal rights and, 155
  consumption of, 50, 55, 58, 66, 69, 81, 99, 120–122, 133, 135, 155, 167, 175
  ecological economics and, 120, 126, 133, 135
  ecotourism and, 167
  First Peoples and, 149
  localization and, 120–122
  permaculture and, 120
Foucault, Michel, 31, 75, 140
Fox, Warwick, 154–155

Fox Keller, Evelyn, 157
Fracking, 19
France, 16, 37, 48, 54
Fuji-Hakone-Izu National Park, 42
Fuller, R. Buckminster, 101

Gaard, Greta, 157
Galapagos Islands, 36, 83
Galbraith, John Kenneth, 61
Gandhi, Mohandas, 62
Gandy, Matthew, 159
Gardens, 6, 25, 28, 36, 38, 64, 66,
    110–111, 121–123, 175, 186n36
Gardiner, Stephen, 88
Gas, 19, 48, 55, 112, 122, 132, 149, 164
Gates, Bill, 88
Gedicks, Al, 147
"Genesis" (Salgado), 42–44
Geoengineering, 71, 84–89, 91, 95–96,
    175
  canals and, 63, 81, 84, 112
  dams and, 26, 78, 81, 84, 99, 119
Georgescu-Roegen, Nicholas, 123
Geothermal energy, 46f, 121, 128
Germany, 6, 12, 36, 49, 51, 54–55, 57,
    59, 125, 132
Ghana, 96
Ghosh, Amitav, 19
Giant (film), 129
Gibbs, Lois Marie, 17
Glaciation, 97
Global Ecovillage Network (GEN),
    119–120
Globalization, 18, 36–38, 62, 105, 118,
    121, 124, 129, 133
Global warming, See Climate change
Golden toad, 36, 78
Goldsworthy, Andrew, 129–130
GRAIL (Gene Recognition and Assembly
    Internet Link), 73
Great Acceleration, 97–99, 98, 113
Great Auk, x, 15
Great Barrier Reef, 15, 45

Great Flood, 104
Great Smog of London, 86
Green accounting, 132, 142
Green buildings, 121–122, 135–136, 168
Green Capital Fund, 136
Greenhouse gases, 49, 55, 87, 99, 136,
    175
Greenhouses, 122, 124
Greenland, 12, 69, 85
Green Party, 67
Greenpeace, 168, 199n4
Greenwashing, 60, 68, 164, 167
Greer, Sue, 17
Griffiths, Tom, 5
Gross domestic product (GDP), 8, 35,
    51–52, 63, 123, 132–133
Grove, Richard, 30
Guatemala, 33
Guattari, Félix, 140
Guha, Ramachandra, 61, 197n31
Gulf of Mexico, 105
Gurin, Sergei, 110–111

Hall, Marcus, 41
Hamilton, Clive, 86–87
Haraway, Donna, 83, 90, 102, 140
Hardin, Garret, 13, 179n27
Hartwig, Atilla, 57
Hawken, Paul, 117–118, 126
Hayles, Katharine, 90
Haynes, Todd, 159
Hedges, Chris, 174
Heidegger, Martin, 101, 129
Heise, Ursula, 29–30, 108
Heller, Lisa, 165
Hengeveld, Rob, 100
Herbicides, 20
Hess, David, 122
Hinduism, 44–45
Hine, Dougald, 106
Holling, C. S., 20–21
Holocene Era, 96
Hooke, Robert, 108

Hornborg, Alf, 104
Houser, Heather, 144
Hulme, Mike, 16
Humboldt, Alexander von, 4
Hunting, 11–15, 19, 23–24, 35, 39–41,
    44, 47, 74, 82, 134, 148–151, 169,
    183nn45,51
Hurricanes, 33, 107, 171–172
Huxley, Aldous, 90
Hydropower, 48, 68, 84, 99, 119

Iceland, 15, 46f, 50
Imagination, 27, 29, 43, 89, 152, 158
India, 19, 29–30, 120, 129
    Bangalore, 66–67
    Bhopal, 105
    Bhutan and, 36
    birth rate, 51, 100
    carbon emissions of, 53
    consumption levels of, 59
    localization and, 120, 122
    Nepal and, 35–36
    religion and, 45
    waste exports to, 18
Indian Ocean, 105
Individualism, 23, 67, 157, 169
Indonesia, 80, 84, 105, 119–120, 124,
    167
Industry, 4–5, 19, 29, 55, 71–78, 84, 88–
    89, 99, 105–106, 113, 132–135, 166,
    167, 172. See also Chemicals, Fossil
    Fuels, Mining, Recycling, Toxins.
    agriculture and, 39, 64, 99, 155
    cancer and, 167, 172, 189n3
    consumption and, 56, 58–59, 124
    corporate images and, 139
    ecotourism and, 31–32, 34, 47, 167
    energy and, 48–49, 52, 55, 104, 168
    First Peoples and, 146–148, 150
    nuclear power and, 34, 48, 105,
        109–111
    oil spills and, 19, 57, 75, 131, 147
    pollution and, 19, 57, 75, 105, 131, 147

queer ecology and, 159
    restructuring knowledge and, 164
    tobacco, 167
Instituto Terra, 42
Intergovernmental Panel on Climate
    Change, 93
International Ecotourism Society, 33
International Geological Congress, 104
International Monetary Fund (IMF), 124
Internet, 5, 14, 41, 58, 73, 90, 91, 167
InterTribal Sinkyone Wilderness Coun-
    cil, 149
Inuit, 150
Inventions, 5, 71, 73, 105, 130
    ecotourism and, 34
    geoengineering and, 87–88
    language and, 10
Iroquois, 25, 176
Isle Royal, 37–39

Jackson, J. B., 24
Jackson, Shelley, 159
Jamie, Kathleen, 29
Jane Goodall Institute, 14
Japan, 41, 45, 51, 53, 62, 68, 124,
    183n51
Jensen, Derek, 77
Jesus Christ, 62–63
Jevon, William Stanley, 4
Jung, Carl, 146–147

Kant, Immanuel, 9
Kathmandu, 35
Kaya Identity, 51–54, 63, 69, 123
Keller, Evelyn Fox, 157
Kelly, Kevin, 10–11
Kelsey, Elin, 107
Kenya, 17, 41, 60, 119, 183n45
Keystone XL Pipeline, 149
Kingsnorth, Paul, 106
Kirksey, Eben, 39
Klein, Naomi, 105, 133–134, 149
Koford, Carl, 79

Kohn, Eduardo, 39, 170
Kolbert, Elizabeth, 14, 102
Krech, Shepard, III, 151, 198n46
Krupa, Jim, 113–114

Laing, Olivia, 29
Lanchester, John, 131
Land Art, 129–130
Landfills, 17, 118, 130, 149, 165, 168,
    173
Landless Worker's Movement (MST),
    120
Land use, 7, 40–41. *See also* Agriculture,
    Forests, Mining
  bayous and, 112
  dams and, 26, 78, 81, 84, 99, 119
  dumping and, 17, 89, 118, 130, 149,
    165, 168, 173
  ecoracism and, 18, 21
  ethics and, 93
  fracking and, 19
  pastures and, 14, 28, 155
  pipelines and, 19, 48, 112, 132, 149,
    164
  place and, 23–24
  psychological effects of, 112, 145–146
  raw materials and, 11, 18, 23, 63, 67,
    118, 124, 127, 156, 167
  sacrifice zones and, 18, 172–174
  Trust for Public Land and, 149
Land Use Database, 129
Language
  animals and, 154
  biodiversity and, 80, 113
  cooption of, 79–80, 113
  environmental disasters and, 100
  numbers as a, 131
  place and, 146
  scientific, 80, 172
Lapland, 33, 89
Latin America, *See* South America.
Latour, Bruno, 140, 143
Lawn, Philip, 133

Lead, 18, 167, 173
Lecain, Tim, 142
Lemenager, Stephanie, 112–113, 129
Leopold, Aldo, 3, 29, 62, 93, 95, 114
Lerner, Steve, 172
Le Roy, Edouard, 99
Lévinas, Emmanuel, 152–153
Linnaeus, Carl, 98
Literature, 104–105. *See also* Ecocritcism
    and individual writers.
Litfin, Karen, 120
Lobbyists, 55, 167
Localization, 118–123, 128–129
  ecological economics and, 133, 135
  internal challenges and, 176
  recycling and, 118, 120–122, 128–129,
    133, 135, 176
  Transition Towns movement and, 122,
    172
Lokuge, Chandani, 34
Long, Richard, 129–130
Lopez, Barry, 26–27, 169
Lousley, Cheryl, 81
Love Canal, 105
Lovins, Amory, 126
Low Carbon City Development Index
    (LCCI), 121
Lufa Farms, 124

Mabey, Richard, 29
MacArthur Workshop on Humanistic
    Studies of the Environment, 3
MacCannell, Dean, 31, 33
MacDonald, Helen, 29
Macfarlane, Robert, 29
Madagascar, 84
Makah tribe, 150
Malaysia, 124
Malm, Andreas, 104
Maniates, Michael, 61
Maori, 82, 89
Margulis, Lynn, 78
Marris, Emma, 38

Marsh, George Perkins, 3–4, 26
Martinez-Alier, Joan, 148
Marxism, 56, 105, 133, 141–142
Masdar City, 66, 68
Mass Extinction Memorial Observatory, 92f, 107–108
Matrix (film series), 143
Mavhunga, Clapperton, 148
Mayr, Ernst, 80
McCalman, Iain, 45
McDonough, William, 126–127
McKibben, Bill, 3, 101–102, 133
McKinsey & Company, 2
McLaren, Duncan, 66
McNeill, John, 97–98
Mead, Margaret, 152
Mech, L. David, 37–38
Melathopoulos, Andony, 86
Melosi, Martin, 17, 174
Merchant, Carolyn, 6, 157
Merton, Thomas, 63
Mexico, 33, 37, 62, 78, 105, 120
Military, 12, 15, 39, 75, 77, 86, 94, 96, 139, 149
Mining, 95–96, 105, 111–113
  Australian Western Mining Corporation and, 17
  biodiversity and, 42, 79, 80
  biotechnology and, 75
  degrowth and, 123, 125, 133
  ecological economics and, 133, 135
  First Peoples and, 147, 149, 197n34
  Great Barrier Reef and, 45
  literature and, 104–105
  pollution and, 48, 105, 173
  space, 75, 187n10
  strip, 48, 99, 113, 148, 174
Modernism, 101, 106, 122, 140, 146
Mollison, Bill, 120
Monteverde Cloud Forest Reserve, 36
Montreal Protocol, 166
Moore, Jason W., 142
Moore, Kathleen D., 3

Moore, Steven, 67–68
Moose, 37–39, 134
Morals. *See* Ethics, Religion, Spiritual values
More, Thomas, 63
Morgan, J. P., 131
Morton, Timothy, 141–142, 157, 158
Mossville, Louisiana, 172
Mountains, 42, 112, 113, 147, 174
Muir, John, 3
Mumford, Lewis, 62, 163
Myers, Natasha, 74–76

NASA, 75
National Academy of Science, 87
National Association of Audubon Societies, 79
National parks, 13, 16, 18, 22, 32, 35–36, 39, 42, 148, 175. *See also* Wilderness
National Research Council, 88
Native Americans, 18–19, 23, 24, 25, 48, 146–150, 154, 176
Naturecultures, 4, 18–19, 21, 175
Nature reserves, 13, 35, 38, 40
Navajo, 24, 183n48
Neolithic revolution, 96–97
Neo-Malthusianism, 61–62
Nepal, 35–36
Nestle, Marion, 58
Netherlands, 39, 59, 61, 63–64, 122, 127, 182n39, 186n43
New Dark Age, 106
New materialism, 126, 140–145, 160, 176
New wilds, 37–45, 150
Nichols, Jeff, 108
Nicholson, Adam, 29
Nigeria, 37, 147
Nixon, Rob, 19, 102, 110, 148
Nobel Prize, 71, 166
Nongovernmental organizations (NGOs), 50, 106, 117, 166–167

Noösphere, 99
Norway, 39, 52
Nuclear energy
    Chernobyl and, 34, 48, 105, 109–111
    consumption and, 48–49, 52
    geoengineering and, 84
    radioactive waste and, 48, 96–97, 109–
        111, 149, 173
    slow violence and, 95
    Three Mile Island and, 48
    weapons and, 8, 84, 100, 105, 109

Oak Ridge National Laboratory, 73
Occupy Wall Street movement, 103, 174
Oceania, 40
Oceans
    acidification of, 84, 97
    Anarctic, 85
    Arctic, 84
    carbon and, 85, 95–97
    cooling of, 121
    fishing of, 130, 135, 150
    food chain and, 112
    Great Barrier Reef and, 15, 45
    Indian, 105
    industry and, 89
    pollution and, 1, 89, 105, 175
    warming, 45
Ohio Key, 162f, 167
Oil, 75–76, 79, 122
    Arctic drilling and, 167
    bayous and, 112
    consumption and, 48–49, 55, 57–58,
        62, 129
    ecological economics and, 131,
        133–134
    Exxon and, 55, 109, 131
    First Peoples and, 19, 147, 149–150
    fracking and, 19
    greenwashing and, 167
    lobbyists and, 55, 167
    OPEC and, 48
    petromelancholia and, 112–113

pipelines and, 19, 48, 112, 132, 149,
        164
    Shell and, 167, 174
    spills and, 19, 57, 75, 105, 109, 131,
        147
    waste and, 70f, 79, 112
Oldenziel, Ruth, 64
Oostvaardersplassen, 39
Operation Popeye, 86
Oreskes, Naomi, 97–98, 103
Organization of the Petroleum Export-
    ing Countries (OPEC), 48
Orr, David, 100–101
Oswald, Alice, 29
Owen, David, 65
Ozone, 166

"Particle Falls" sign, 128
Pastures, 14, 28, 155
Patents, 71, 75, 77, 88
Pellow, David, 173
People Against Hazardous Waste Land-
    fill Sites, 17
Permaculture, 120–123
Peru, 147
Pesticides, 16, 19, 26, 100, 167, 171
Peterson, Rolf, 37–38
Petrochemicals, 17, 172–173
Philippines, 17, 124
Phillips, Dana, 101
Phipps Conservatory and Botanical Gar-
    dens, 128
Picher, Oklahoma, 172–173
Piketty, Thomas, 125
Pipelines, 19, 48, 112, 132, 149, 164
Place, 23–32, 175
    biodiversity and, 25, 29, 31, 35, 38, 45
    biotechnology and, 71–78
    ecotourism and, 30–37, 44
    extinction and, 24, 36, 38, 41
    new wilds and, 37–45
    psychology of, 24, 27, 40–41, 111–
        112, 145–146

resources and, 27, 29, 32, 35, 38,
    40–43
sacred places and, 19, 24, 29, 44, 147,
    149
Thoreau and, 27–30, 42
wilderness and, 24, 29, 37–45
Plantations, 15–16, 98, 174
"Play the LA River" (Project 51), 116f,
    130
Plumwood, Val, 153–154, 170
Pollan, Michael, 58
Polli, Andrea, 128
Pollution, 17–19, 26, 48–52, 64–65,
    75–76, 105, 109–111, 134, 144, 148–
    149, 156, 168, 172–173
acids and, 8, 18, 48, 84, 97, 102
air, 1, 35, 51, 55, 57, 86, 105, 128, 173
cancer and, 17, 109–111, 167, 172,
    189n3
chemicals and, 99, 105, 166, 172–173
degrowth and, 124–126
ecotourism and, 35
environmental humanities and, 1, 3,
    7–8, 13, 17–19
hurricanes and, 33, 107, 171–172
industrial, 19, 57, 75, 105, 131, 147
oceans and, 1, 89, 105, 175
oil spills and, 19, 57, 75, 131, 147
recycling and, 128
restructuring knowledge and, 164
solar energy and, 55, 60
toxic, 17–18, 75, 105, 109, 134, 144,
    148–149, 156, 168, 172–173
water, 7–8, 17, 19, 35, 55, 172–173
Pope Francis, 10
Population, 26, 28, 36–39, 51–54, 100,
    158. See also Consumption, Kaya
    identity
Portugal, 55, 120
Postcolonial studies, 4, 6, 8, 18–19, 34,
    140, 145–147
Posthumanism, 4, 11, 78, 89–91, 102,
    140, 145, 152, 159, 175

Postmodernism, 140
Poverty, 136, 150
Preservation
    animals and, 13–19, 22, 32, 35–36, 39,
        42, 148, 175
    biodiversity and, 25, 29, 80, 83
    ecotourism and, 31, 34–37
    First Peoples and, 146, 151
    national parks and, 13, 16, 18, 22, 32,
        35–36, 39, 42, 148, 175
    nature reserves and, 13, 35, 38, 40
    wilderness and, 17, 29, 37–38, 44–45,
        159, 175
Price, Jenny, 106, 129
Princen, Thomas, 61, 118–119
Project 51, 116f, 130
Project Zero-Bright Green Business, 121
Psychology, 105, 111–112, 145–146
    identity and, 24, 27, 40–41, 63
    place and, 24, 27, 40–41, 111–112,
        145–146
Purdy, Jedediah, 87

Quaternary Period, 96
Queer ecology, 3, 156–160
    Abney Park and, 159
    animals and, 152, 160
    anthropology and, 157
    consciousness and, 176
    environmental justice and, 157–159
    evolution and, 78
    feminism and, 157–158

Radioactive waste, 48, 96–97, 109–111,
    149, 173
Rain forests, 24, 80–82, 93, 106, 135
Raw materials, 11, 18, 23, 63, 67, 118,
    124, 127, 156, 167
Recycling, 59, 65, 123–129, 139, 165,
    175
    carbon footprint reduction and, 20
    ecological economics and, 133, 135
    ecotourism and, 33, 37

Recycling (cont.)
  localization and, 118, 120–122,
    128–129, 133, 135, 176
  neo-apocalyptic alternatives and, 117
  upcycling and, 62, 127–128, 167
  water and, 121–122, 127–128
Reece, Erik, 113–114
Reforestation, 36, 39, 42, 83, 98
Refrigerators, 10, 51, 54–55, 64, 166
Regan, Tom, 154
Regulation
  biotechnology and, 73
  building codes and, 65, 68
  Convention on Biological Diversity
    (CBD) and, 79–80
  EU "End of Life Vehicle Directive"
    and, 127
  geoengineering and, 86
  industry and, 17, 109
  local communities and, 174
  sacrifice zones and, 18, 172–174
  toxins and, 109, 144, 172
  Wilderness Act and, 40–41, 183n43
Religion
  animals and, 14, 41, 152
  biodiversity and, 80
  biotechnology and, 73
  Buddhism, 11, 35, 45, 62–63
  Christianity, 10–11, 40, 43, 62–63,
    184n55
  First Peoples and, 11, 147, 154
  Hinduism, 44–45
  materialism and, 62–63
  monks and, 62–63
  rituals and, 24, 62, 75, 82, 145, 170
  sacred places and, 19, 24, 29, 44, 45,
    82, 147, 149–150
  spiritual values and, 27, 42, 45, 120,
    141–142, 150, 154
Renewable energy, 28, 50, 55, 60, 66, 68,
  121–122, 128, 131, 133, 136, 166–167
Renewable Energy Buyers Alliance,
  166–167

Resilience, 7, 20–21, 122
Resources, 12–14, 17–18, 52, 59, 66,
  118. *See also* Consumption, Energy,
    Sustainabilty
  Anthropocene and, 95, 100–101
  biodiversity and, 80, 84
  degrowth and, 124, 126
  ecological economics and, 132–133, 135
  First Peoples and, 147–152, 166
  place and, 27, 29, 32, 35, 38, 40–43
  recycling and, 126–129
  World Resources Institute and, 166
Revkin, Andrew, 102
Rituals, 24, 62, 75, 82, 145, 170
Robin, Libby, 8
Robinson, Florence, 172
Robinson, Kim Stanley, 89
Robinson, Marilynne, 147
Robinson Forest, Kentucky, 113
Rocky Mountain Institute, 166
Rodney, Walter, 147
Royal Society, 87
Russia, 48, 54, 84, 109–111

Sacred places, 19, 24, 29, 44, 82, 147,
  149–150
Sacrifice zones, 18, 172–174
Sahara Desert, 85, 175
Saint Francis of Assisi, 10
Saint Helena, 30
Salgado, Sebastião, 42–44
Sally Bell Grove, 149
Salmon, 20, 78, 81–82, 155, 165
Salmon People, 82
Samoa, 89
Samsø, 122, 123
Sandhill cranes, 83
Sandilands, Catriona, 157–158
Sanitation, 35, 174
Sarvodaya project, 120
Schulke, Flip, 162f
Schumacher, E. F., 61, 120
Scott, Slovic, 17

Scranton, Roy, 94
Sekula, Alan, 129–131
Senegal, 89
Sennett, Richard, 67
Serpas, Martha, 112
Seymour, Nicole, 158–159
Sheep, 14, 56, 120, 129
Shell Oil Corporation, 167, 174
Silent Spring (Carson), 26, 63, 105, 109
Silko, Leslie Marmon, 147
Sinclair, Cameron, 134
Singer, Peter, 155
Sioux, 11, 19, 149, 154
Skocpol, Theda, 55
Slavery, 15–16, 135, 155, 174
Sloterdijk, Peter, 101
Slow Food, 167
Slow violence, 19, 21, 95, 112, 148, 171, 175
Smithson, Robert, 130
Snyder, Gary, 44
Sobecka, Karolina, 89
Soil, 75, 97, 110–113, 155, 168, 172. *See also* Agriculture, Place
    crop yields and, 106
    erosion and, 100, 113
    fertilizers and, 84–85, 89, 155
    First Peoples and, 39
Solar energy,
    building design and, 28, 68, 128, 132
    carbon footprint and, 52
    ecological economics and, 50, 132, 136
    ecotourism and, 32, 36
    efficiency of, 55
    localization and, 121–122
    Masdar City and, 66, 68
    stoves and, 60
Solar radiation management (SRM), 85–86, 89
Solastalgia, 112
Sønderborg, 65, 121
Sörgel, Herman, 84
Sörlin, Sverker, 4, 103, 169

South Africa, 18, 96, 112, 178n13
South America, 17, 39, 41–44, 81, 84, 119–121, 132, 147, 166, 173. *See* specific countries and Amazonia
    Columbian exchange and, 98
    ecological imperialism and, 15–16
    rights of nature and, 173–174
Soviet Union, 34, 48, 54, 84, 105, 109–111
Soylent Green (film), 105
Spiritual values, 27, 42, 45, 120, 141–142, 150
Sri Lanka, 34, 119–121
Star Wars (film series), 90
Stasch, Rupert, 43
Stein, Rachel, 149
Stengers, Isabelle, 60–61, 83
Stiglitz, Joseph, 125
Stoermer, Eugene, 95
Stohr, Kate, 134
Stone Korowai, 43
Stoner, Alexander, 86
Stoppani, Antonio, 98
Stratigraphic Commission, 96–98
Subra, Wilma, 172
Suess, Eduard, 99
Sulfur, 5, 75, 85–86, 89
Sullivan, Robert, 29
Sunlight Reflection Methods, 87
Sustainability, 1, 12–14, 20–21, 35, 47, 50–69, 117–118, 120–127, 164, 175. *See also* Cities
    Anthropocene and, 95, 100, 106
    biotechnology and, 77
    as corporate slogan, 139
    ecological economics and, 132–133, 136, 176
    ecotourism and, 32–35, 167
    First Peoples and, 149
    hunger and, 106
    military and, 139
    new materialism and, 144
    restructuring knowledge and, 164

Suzuki, David, 117
Sweden, 6, 41, 65, 103, 111, 119–120
Switzerland, 32, 59

Tagebau Garzweiler mine, 125
Take Shelter (film), 108–109
Tall, Deborah, 25
TallBear, Kimberly, 151
Tanner, James, 79
Tanzania, 12, 14, 17, 34, 41, 183n45
Taxes, 52, 124–125, 133, 168
Tennessee Valley Authority, 119
Terminator 2 (film), 130
Thoreau, Henry David, 3, 27–30, 42,
    61–62, 81–82, 158
Tibet, 19, 44, 62
Tobacco industry, 167
Toxins, 105, 134, 149–150, 172–173
  biotechnology and, 75
  dumping and, 17, 89, 118, 130, 149,
    165, 168, 173
  exporting of, 18, 62, 112
  food and, 109, 134
  hazardous waste and, 17
  lead and, 18, 167, 173
  new materialism and, 144
  regulation and, 109, 144, 172
  transcorporeality and, 156
Tragedy of the commons, 12–13, 21,
    168, 23, 175
Transcorporeality, 156–157
Transgenic species, 77, 156
Transition Towns movement, 122, 172
Trinity Test, 96
Trust for Public Land, 149
Tuan, Yi-Fu, 24
Tucker, Cora, 17

UNESCO, 36, 42, 123
Union of Geological Sciences' Subcom-
    mission on Quaternary Stratigraphy,
    96–98

United Nations
  Conference on Environment and
    Development and, 81
  Convention on Biological Diversity
    (CBD) and, 79–80
  ecotourism and, 35
  Goal Thirteen and, 136
  Indigenous People's Global Summit
    on Climate Change and, 89
United States, 18, 19, 65, 71, 86, 120,
    124–125. *See also* individuals by
    name.
  activism, 130, 174
  African Americans, 17, 149
  climate change and, 131, 176
  consumption and, 49–53, 56–57, 59,
    65, 67, 79–80, 112, 132
  corporate green initiatives and, 32,
    166–167
  demand for green buildings and, 66,
    135–136
  emergence of environmental humani-
    ties and, 3–4, 6, 105, 157
  environmental justice movement and,
    17, 107, 149, 172–173
  forests, 39, 74, 98, 113, 149, 150
  government and environment of, 119,
    131, 135
  Native Americans, 11, 18, 23–25, 48,
    146, 147, 149, 150–151, 154, 176
  pollution, 18, 48–49, 65, 78, 109, 124,
    128, 172
  sense of place and, 6, 12, 23–26, 39,
    130
  utopian communities, 121
  wilderness concept and, 18, 39–41
Universal Declaration of Human Rights,
    102
Universal Declaration of the Rights of
    Nature, 174
Upcycling, 62, 126–128, 167, 174
Upper Xingu Basin, 43

Urry, John, 31–32
Utopia (More), 63

Van Dooren, Thom, 83
Venezuela, 44
Verkerk, Mark, 40, 182n39
Vernadsky, Vladimir, 4, 99
Vildmark, 41
Violence, 19, 42, 83, 106, 147, See also
   Slow violence
Volkswagen, 57
Von Mossner, Alexa Weik, 108

Walden (Thoreau), 3, 27–29
Walking Dead, The (TV series), 93–94
Wane, Njoki, 59–60
Wanick, Lelia Deluiz, 42
Waste, 7, 13, 17, 18, 32–33, 48–51, 66,
   118–124
  Anthropocene and, 103, 105, 112
  biodiversity and, 79
  ecological economics and, 133, 139
  e-waste and, 18, 62, 112, 166
  First Peoples and, 149
  food, 110
  new materialism and, 144
  nuclear, 48–49, 96–97, 109–111, 149,
   173
  oil, 19, 57, 70f, 75, 79, 105, 109, 112,
   131, 147
  recycling and, 127–129, 175
  sanitation and, 35, 174
  sustainable cities and, 64–66
  toxic, 17–18, 75, 105, 109, 134, 144,
   148–149, 156, 168, 172–173
Water, 17, 19, 20, 26, 33, 45, 72, 93, 97
   100, 112, 119, 171
  biomining and, 75
  canals and, 63, 81, 84, 112
  dams and, 26, 78, 81, 84, 99, 119
  degrowth and, 126, 169
  ecological economics and, 134, 136

ecotourism and, 33, 35
First Peoples and, 24, 44, 148–149
geoengineering and, 84–85
pollution of, 17, 19, 35, 55, 134,
  171–173
recycling and, 121–122, 127–128
sustainable cities and, 64, 171
Watts, Alan, 9
Weapons, 12, 75, 77, 86, 96, 149
Weisman, Alan, 103
West Nile virus, 105
West Papua, 43
Wetlands, 39, 44, 76, 112, 117, 173
Weyler, Rex, 168
Whales, 35, 82, 149–150, 154
Wheeler, Sarah, 29
White, Richard, 129
Whitman, Walt, 153
Wilderness, 15–18, 21, 22, 24, 29,
  37–45, 159, 175. See National parks
  Amazonia and, 16, 39, 43, 98, 150,
   171
  American West and, 15, 19, 39
  as bankrupt concept, 169
  biotechnology and, 76–77
  commodification and, 130
  environmental justice and, 17
  First Peoples and, 24, 98, 149
  InterTribal Sinkyone Wilderness
   Council and, 149
  new wilds and, 37–45, 150
  queer ecology and, 157
Wilderness Act, 40–41, 183n43
Williams, Raymond, 141
Williams, Terry Tempest, 3, 109
Wilson, E. O., 81, 107, 128
Wind energy, 121
  Amazon and, 166
  building design and, 28, 128, 132
  carbon footprint and, 52
  ecological economics and, 50, 132
  First Peoples and, 62

Wind energy (cont.)
  localization and, 122
  pollution and, 55
Wolves, 21, 37–38, 41, 169, 183n51
Wooley, Agnes, 108
World Bank, 60, 119, 124, 132
World Resources Institute, 166
World Wildlife Fund (WWF), 121,
    166–167

YouTube, 167

Zalasiewicz, Jan, 96
Zimbabwe, 148
Žizek, Slavoj, 102
Zoos, 153